D1066612

Guide to Electronic Surveillance Devices

by Carl Bergquist

THOMSON

DELMAR LEARNING

Australia • Canada • Mexico • Singapore • Spain • United Kingdom • United States

Guide to Electronic Surveillance Devices
by Carl Bergquist

Business Unit Director:
Alar Elken

Executive Editor:
Sandy Clark

Senior Acquisitions Editor:
Gregory L. Clayton

Editorial Assistant:
Jennifer Luck

Executive Marketing Manager:
Maura Theriault

Channel Manager:
Mary Johnson

Marketing Coordinator:
Karen Smith

Executive Production Manager:
Mary Ellen Black
Production Manager:
Andrew Crouth

Production Coordinator:
Sharon Popson

COPYRIGHT 2002 by Delmar Learning, a division of Thomson Learning, Inc. Thomson Learning™ is a trademark used herein under license.

Printed in Canada

1 2 3 4 5 XX 05 04 03 02 01

For more information contact Delmar
5 Maxwell Drive
Clifton Park, N.Y. 12065

Or find us on the World Wide Web at
http://www.delmar.com

ALL RIGHTS RESERVED.
No part of this work covered by the copyright hereon may be reproduced in any form or by any means—graphic, electronic, or mechanical, including photocopying, recording, taping, Web distribution, or information storage and retrieval systems—without the written permission of the publisher.

For permission to use material from the text or product, contact us by
Tel.   (800) 730-2214
Fax   (800) 730-2215
www.thomsonrights.com

Library of Congress Cataloging-in-Publication Data:

LCCN 2002103404

Guide to Electronic Surveillance Devices / Carl Bergquist

ISBN: 0-7906-1245-3

Contents

PART I: BACKGROUND

PART II: AUDIO SURVEILLANCE

PART III: VIDEO SURVEILLANCE

PART IV: SPECIAL ISSUES

Acknowledgments

The author would like to thank Steve and Jake at Supercircuits for their assistance and the use of several of the excellent photographs from their catalog.

A book like this doesn't get published without the hard work of a number of people in addition to the author. Project managers, technical reviewers, compositors, proofreaders, and, of course, editors have contributed many hours of effort (and sleepless nights) to whip this text into shape. I would like to thank the entire project team at Abshier House and Delmar Learning for their contributions to this book.

Thanks, guys!

Carl

About the Author

Following a successful career as a photojournalist for the AP, UPI, *The New York Times*, *Newsweek*, and other publications, author Carl Bergquist turned his efforts towards his lifelong hobby of electronics. He has published articles in *Popular Electronics*, *Electronics Now*, and other electronics related magazines, and written a series of books for Prompt Publications.

Trained in chemistry and history, circumstances led him into the world of photojournalism, where he remained for over twenty years. Today, in addition to his work with electronics, Mr. Bergquist writes for a U.S. Air Force newspaper. His other interests include golf, amateur radio, computer technology, gemology, the American Civil War, and another of his lifelong loves, flying. However, writing and photography occupy much of his time.

Technical Editors

Nancy E. Murphy has been with the world's largest provider of electronic security services for twenty-five years. She has worked in all of the technical facets of this industry, including installation, repair and service of CCTV, and burglary/fire and access control within the residential, business, and commercial aspects. Today, she is a district service manager overseeing the day-to-day operations and work of forty service technicians that provide repair and service to a customer base of over 60,000 customers.

Patrick D. Zakaria has been in the electronic security industry since 1994. He received extensive training in the field of communication electronics while enlisted in the United States Marine Corps and continued his education in the field of CCTV, burglary/fire and access control. He is NICET certified. He is currently the maintenance team manager of twenty-seven technicians who service and maintain high-end integrated security systems for the world's largest provider of electronic security. He oversees service for a customer base of 30,000.

Introduction

Electronic surveillance has come a long way in just the last ten years, and today it is easier than ever to protect both your home and business through the use of surveillance devices. While audio observation is still employed to some extent, when I say *surveillance* I really refer to video or pictures. In this realm lies the ability to not only hear what is going on but be able to see it as well. The latter offers a great advantage in how well you actually do protect your property, employees, friends, family, and yourself.

This text is written for anyone who has the need to observe activity in or around his or her home or business. In reality, that includes just about all of us. It might not be necessary to maintain twenty-four-hour surveillance in all circumstances, but there are going to be times when the ability to see what is happening in certain areas of the home/business will be reassuring, pragmatic, or both.

Naturally, safety is often a prime factor in installing a surveillance system. In both the home and business environment, dangerous situations, volatile interactions, or menacing locations can be watched and/or resolved through the use of surveillance gear. In the long run, this equipment will usually pay for itself many times over in just peace of mind alone. To that extent, home and business surveillance is well worth looking into.

In recent years, video cameras have experienced a renaissance that is nothing short of miraculous. A circuit that used to require an eight by six by four-inch case can now be put on a printed circuit board (PCB) not much larger than a standard first-class postage stamp (but thicker, of course). By the way, we are talking about both black-and-white and color video cameras. What this means to you is the availability of convenient, inexpensive, and far superior video devices that can help you keep a vigil on your home and business.

If you have a need for video surveillance, there has never been a better time to venture into this field. Not only has equipment size been reduced to an almost unbelievable state, but price has shrunk right along with size. You can install a very sophisticated system for what a single camera arrangement would have set you back ten years ago. I well remember the first video gear I bought. Cameras were in the $500 range, monitors were expensive, bulky, and unreliable, and power requirements were ridiculous (especially hard on the power bill).

In this text, we will explore virtually every aspect of electronic surveillance. I have made it a point to cover every type of equipment that exists as of the writing of this book. But this is an ever-changing field that always has new surprises in store.

We will look at the various cameras, monitoring equipment, and image-recording devices available to the consumer. We will also delve into the methods of getting the picture from the camera to the monitor. For those of you more curious in nature (like me), I will discuss some of the really "gee-whiz" stuff that abounds out there. I must admit this part always captivates me. I'm like a kid in a toy store.

Naturally, our journey through the world of electronic surveillance will take us into the areas of installation and use of the equipment. And, for the convenience and protection of the reader, I will cover some of the pitfalls involved in this aspect of surveillance. I will walk you through the equipment, explain how to place cameras for optimal viewing, and describe how to use the gear to its best advantage. I will also touch on audio surveillance and some of the things that can be done there. And, for the heck of it, I will provide a short history of video and video surveillance and a technical, or "how it works", approach to the equipment.

> **Note:** When considering your options for equipment and techniques, remember that this industry, like every other technical field, is experiencing almost daily changes. Keeping pace with the latest gadgetry is virtually impossible unless your time and budget are unlimited, and this text has experienced the same problem. This book shows and discusses the latest equipment generally available at the time of publication, but the industry keeps evolving and new techniques come to light practically every day. When making selections, be sure to consult a qualified professional (such as those listed in Appendix B) for assistance.

PART ONE

Background

*Tempted to skip this part of the book and go directly to the "good stuff"? Try to restrain yourself. Although knowing some of the background behind the development of surveillance equipment is not a requirement before you can set up a security system, it can be extremely helpful to grasp the concepts behind the basic equipment technologies. And while you don't necessarily need to understand all the bits and bytes of how this electronic equipment works, you do need at least some comprehension of how not to end up in jail. Yes, **jail**. There are legal ramifications when setting up even a simple system to guard the perimeter of your personal property. The chapters in this part of the book explore some of these basic issues and provide important pointers to areas you must consider before setting up your system.*

1

A Brief History of Electronic Surveillance

Surveillance in the twenty-first century makes use of a highly sophisticated collection of devices that are nothing short of miraculous. However, it wasn't always that way. And how we got to where we are is a story of man's ingenuity, knowledge, and adaptability. But before we launch into how to select and set up equipment, let me provide a brief history of electronic surveillance. (It won't be all there is to know about the subject; when I say *brief*, I mean *brief*. But I hope you'll find it as fascinating as I do.)

The term *electronic* would generally be considered to include voice-only systems or radio. However, most of what we know today as *electronic surveillance* involves *seeing* what's happening in a given area. That, of course, involves video or television. So most of our history will cover the events that led to modern-day video. I'll discuss radio, of course, but I'll leave the details on its history to other books. Instead, most of this chapter will focus on the development of video technology and television.

Fascinated yet?

Without further fanfare, let's jump into history.

HOW RADIO STARTED IT ALL

With the invention of radio came a whole new concept in transferring information. Early radio employed Morse code only and was of no use for surveillance purposes. When voice modulation was perfected, however, spies really had something to work with.

Hardwired microphones had been used for many years for bugging, but such microphones and their associated cables were difficult to conceal, and thus easy to spot. Additionally, setting up hidden microphones was a laborious and time-consuming task at best. Hence, their use was somewhat limited.

With radio to replace those cumbersome and all-too-often obvious cables, such listening devices could be installed faster and more easily. The practice of using microphones for clandestine surveillance became more prevalent, and those who benefited had radio to thank. The microphones were still big and hard to hide, but at least you didn't have to hide all that wire as well.

Naturally, once this application of radio became known, there was some disapproval of the practice. Oddly, it was many years before laws were passed making electronic bugging illegal, but that was not for lack of complaints about wireless listening, especially when those whose activities were subject to sound surveillance couldn't detect that they were being listened to.

Of course, to this day covert audio listening devices are used frequently. Corporate eavesdropping has become a significant issue—so much so that individuals with the equipment and skill to detect and remove such bugs can make a darn good living.

Be that as it may, laws now exist prohibiting audio surveillance—in some cases, even unintentional audio surveillance. In the state of Florida, for example, using any device that listens to or records audio is a felony, and can bring a long stretch in the "ol' stony lonesome" (jail). In a recent case, someone was secretly videotaping activities in a room. Because he forgot that the camcorder also recorded audio, he ended up doing hard time. Actually, it wouldn't have made any difference if he had remembered the audio, though, because he was unaware that in Florida videotaping is okay, but audiotaping is considered wiretapping.

Although the average citizen may generally consider videotaping to be more "serious" than audiotaping, stringent regulations such as those in Florida may apply. If you plan to do any audio surveillance, you must be careful to check local and state regulations regarding this practice. You don't want to run into any surprises, like waking up to a guy in a blue uniform reading you your rights.(Because regulation of surveillance activities has become such a serious issue, I've devoted an entire chapter in this book to that topic. See Chapter 3, "Legal Ramifications.")

Obviously, I could go into great detail on the invention of radio and its lengthy history, both as entertainment and as a surveillance medium. Certainly radio has played an important surveillance role in politics and during war, and is still used in those contexts in many areas today. But far more important in this context is that radio set the stage for what was to come: video surveillance!

ALONG COMES THE MAGIC EYE—TELEVISION

In order to do justice to the invention of television, it must be mentioned at least in passing that the roots of TV go back centuries. I don't want to dwell on this aspect, but the characteristics of light-sensitive chemical elements were understood even in the early 1800s. It was just that nobody saw any use in it until about 1870, when the first *photocell* was invented. Even then, the photocell was considered little more than a "gee whiz" device.

About this same time (1878), a British scientist, Sir William Crooke, developed the first cathode-ray tube (CRT), which he called the *Crooke tube*.

Without realizing it, the inventors of the photocell and CRT had laid the groundwork for modern television. Naturally, there was a lot more to it than just these two electronic components, but without those initial inventions we might have an entirely different type of television today, one that most of us probably wouldn't like very well. This alternative method involved synchronized spinning disks that had little holes in them. As the disk spun, light would fall on the subject through the holes. At the other end, light would be projected through supposedly identical holes to reproduce the original image. The system was a mechanical nightmare that actually endured for more than 50 years before it was laid to rest in favor of electronic television.

The mechanical spinning-disk version of television was the brainchild of German scientist Paul Gotlieb Nipkow. Nipkow placed tremendous faith in his system—unfortunately for him, nobody else did. Actually, there was a very good reason that mechanical TV didn't garner much respect. It didn't work very well, when it worked at all.

However, we do have to give credit where credit is due. Paul Nipkow did demonstrate the basic principle of image scanning that is the backbone of modern television. Without his early system, as crude as it was, there is at least a possibility that electronic television would never have been perfected or would have been a longer time in coming. (Then again, maybe not.)

The next link in the chain of events that made electronic TV possible came in the person of Philo Taylor Farnsworth. The story goes that Farnsworth was out plowing his father's wheat field, or some sort of field. As he plowed each row, it occurred to him that if he could introduce information (shades of gray) in those rows, he would end up with a picture. Philo was so intrigued with this theory that he discussed it with his physics teacher, who encouraged him to seek a viable working system. Philo was also aware of the Crooke tube and decided that the cathode-ray tube would be just the electronic contrivance he needed to test his theory. The first form of electronic scanning was born.

However, Farnsworth had company in his genius. A Russian immigrant named Vladimir Kosma Zworykin had been pondering virtually the same approach to sending pictures over a wire or out into the air. Zworykin had arrived in the United States a few years earlier and had taken a job with George Westinghouse. In 1923, he attempted to patent an imaging tube he called the *iconoscope*, which was a cumbersome device that rather resembled a glass saucepan with a bent handle. The iconoscope used a matrix of tiny photocells as the light-sensitive area and bombarded it with a stream of electrons from an electron gun in the bent-handle part. The image that fell on the matrix regulated the output of the tube and said output was an electrical representation of the image.

There was only one really major problem with the iconoscope: It didn't work! While the iconoscope was sound in theory, Zworykin just hadn't designed the tube properly. Hence, his patent was denied, but a quirk in the law would later come back to haunt Farnsworth and, some say, unfairly benefit Zworykin.

By about 1925, Farnsworth had developed a working image tube that he lovingly referred to as the *image dissector*. Combined with a crude picture tube fashioned from a Crooke tube, this gave Farnsworth the first working completely electronic video system.

Meanwhile, American scientist Charles Francis Jenkins and British scientist John Logie Baird were individually toying with the concept of sending revolving disk pictures via radio waves. While their successes were hardly impressive, the work did advance the thinking along the lines of wireless television.

Zworykin was now employed by Radio Corporation of America (RCA). His boss, David Sarnoff, was convinced that Vladimir was the answer to his dream and obsession. That dream was for RCA to dominate television. Actually, Sarnoff didn't have much competition for that domination, as most industry giants felt that television was at best a neat novelty and at worst a complete waste of time and resources. A well-known corporate honcho of the time, who shall remain nameless (the money he lost due to his lack of foresight was embarrassing enough), was heard to say something like this: "Nobody is going to give up radio for that thing [television]." Needless to say, in the final analysis company stock-holders were not amused!

Sarnoff was pretty much out there alone in his belief that television was going to be BIG. But that didn't slow him down when it came to making RCA the leader in the field. A couple of problems stood in the way of his quest, however. A working electronic system did exist, but it belonged to Philo Farnsworth. Additionally, Vladimir Zworykin was unable to perfect the iconoscope. Sarnoff needed Farnsworth's image dissector, but there was some "bad blood" between the two men, and Sarnoff didn't really want to pay Farnsworth what the device was worth. (Well, in all fairness, RCA wouldn't give him the money.) So he did the next best thing. Around 1930, Sarnoff had Zworykin, masquerading as a Westinghouse engineer, pay Farnsworth a visit at his San Francisco laboratory.

During the sojourn, Farnsworth, who had nothing against Westinghouse, eagerly demonstrated his image dissector. Zworykin was impressed. The presentation must have illustrated the flaws in the iconoscope, because it wasn't long before Zworykin had a working tube.

Sarnoff now had what he wanted: a fully electronic television system. Farnsworth still owned patents on some of the essential systems, requiring Sarnoff to pay royalties, but Sarnoff now had the leverage he needed to get a better television commitment out of RCA.

The iconoscope was problematic in many ways. It required a tremendous amount of light—which was known to set indoor sets on fire and melt actors' makeup. Also, it didn't quite see light the way you and I see it. Thus, people who appeared on television had to wear some very bizarre makeup to keep from looking like Casper the Ghost.

But it did work, and that was all Sarnoff was concerned about, at least for the moment. His faith in Zworykin's ability to produce better image tubes was well founded; in October 1945, Zworykin patented the far-superior *image orthicon* tube. This device would revolutionize television and is largely responsible for the popularity TV would come to enjoy.

Throughout the 1930s, RCA and others operated a number of experimental television stations. Some of the early ones used revolving-disk technology, but all of them transferred to electronics when the iconoscope became available. All of this helped television, still in its infancy, to gain the attention it needed to promote what would become one of the wealthiest industries in recorded history.

World War II put a real crimp in the development of TV. As the nation's focus shifted to the war, virtually every American industry geared toward supporting the Unites States' effort in the conflict. One result was that very little additional research was conducted in the field of video until after the hostilities ended. RCA kept Zworykin busy, however. He continued his work on image tubes, and, as has been noted, perfected the all-important orthicon tube during the war years.

Following World War II, television came out of the chute with a vengeance. In 1946, the first mass-produced and affordable television sets hit the market—and found surprising popularity, considering the limited programming that was available. The end of the 1940s saw far better and more frequent programming, and television sets were becoming commonplace in American homes.

The year 1952 witnessed another milestone in video history. That year the smaller, more light-sensitive, more reliable, and cheaper *vidicon* tube was unveiled by, you guessed it, RCA and company (Vladimir Zworykin). At first used primarily for closed-circuit television systems (CCTV), by the mid-1970s significant improvements in the tube allowed the vidicon to replace the trusted orthicon as the broadcast standard.

The 1970s also marked the arrival of a device that still dominates television and video. The completely solid-state charge-coupled device (CCD) hit the market running and today has become the imager of choice for even broadcast-quality studio cameras. The advantages that the CCD offered over conventional tube-image devices were immediately apparent. Semiconductor reliability, low voltage and current requirements, and minimal heat output generated were just a few of the obvious benefits. Ultimately, lower cost, smaller size, and better performance would join the list of improvements that placed the CCD in the revered position it enjoys today.

That brings us almost to the present in terms of modern video equipment. The CCD and Complimentary Metal Oxide Semiconductor (CMOS) imagers are being improved on a daily basis. Every new batch that manufacturers produce seems to be smaller, less power-hungry and more light-sensitive than its predecessors. And there seems little sign that this trend is going to change in the foreseeable future. See Chapter 8, "Advanced Video Equipment and Accessories," for more information about CCD and CMOS imagers.

WHAT DOES ALL THIS HAVE TO DO WITH MY SURVEILLANCE SYSTEM?

You could do a crack job of setting up your surveillance equipment without any of the information I've dispensed so far (though I hope you've found it interesting). Now I'll move on to discuss how and when video cameras were first employed for surveillance purposes, and how we got to where we are today regarding home and business video protection. As soon as the first "portable" television cameras became available, some people envisioned using them for clandestine purposes. However, the original devices were large and somewhat temperamental. They also used huge amounts of power, generated a lot of heat, were generally hard to hide, and were prohibitively expensive.

This didn't deter everyone. If these obstacles could be overcome, especially the cost, closed-circuit television could be a highly useful tool in a number of scenarios. Under circumstances where visual surveillance could protect property and people or save lives, the impediments were handled and the systems were installed. Sometimes this included special cooling equipment to keep the vacuum-tube-based gear from burning out or burning up. And it might mean renovating a building to create space to hide the cameras. But for applications such as monitoring banks, watching intensive care patients, and aiding in law enforcement, the benefits often outweighed the cost.

Naturally there were people who used the cameras with, shall we say, less than honorable intent. But the early cameras were hard enough to hide in your own surroundings; it was nearly impossible to install them in someone else's. This did deter a lot of what might otherwise have been heavy visual spying. Individuals pursuing industrial espionage would surely have made better use of television if they could have employed these technologies efficiently.

By the mid-1960s, semiconductors and newer manufacturing techniques had begun to shrink both the size of cameras and their price tags. For the first time, vidicon black-and-white units about the size of a small shoebox were available to the general public in the $200 range. Some of these were all tube gear, employing miniature vacuum tubes, but others had begun to use transistors for some of the circuitry.

Now placement of TV cameras for either overt or covert purposes became a much easier matter, and video surveillance experienced a sudden growth that made a lot of people very happy. Not only could homeowners and businessmen help protect their own safety and income, but manufacturers of the equipment were getting rich selling cameras and monitors. Many people were indeed very pleased with the newer, smaller video gear.

Sadly, less-virtuous types also found ways for these cameras to perform unsavory tasks, but that's always going to be the case. Overall, the good accomplished by reputable surveillance often counteracted the bad done by the crooks.

Today, video surveillance of either type has come to an almost ridiculous stage. Cameras have shrunk almost to the size of postage stamps, and wireless transmitters are about as small. The largest component of many wireless video systems is often the battery or the antenna. And this is color video, not just black-and-white. Size and power requirements are no longer deterrents to setting up a very viable video surveillance arrangement. Furthermore, the cool-running characteristics of semiconductors, reduced cost, and general reliability of this gear enhance the prospect of being able to watch various areas of your home and business on television screens.

CONCLUSION

In the final analysis, technology in the form of solid-state imagers, integrated circuits, and individual components has brought the concept of electronic surveillance to a new high. Today, it is easier than ever to buy, install, and reap the benefits of this extraordinary video equipment. If you need to watch your property, this may be the answer.

2

Purposes of Home and Business Surveillance

Other than recording, um, private activities, what are some of the reasons you might want to put video surveillance equipment in your home or business? Naturally, video cameras allow you to monitor activity no matter where they're placed, and that has some very pragmatic applications. But there are some other incentives that might induce you to install viewing gear. This chapter explores some of those motives.

SECURITY

There are many reasons for surveying your property, and security, or keeping your home and business safe from undesirable individuals, is probably the most obvious. When talking with people about this book, the first purpose to come to mind for most people was the ability to keep an eye on their houses or places of business. Thus, security ranks high on the list.

When you can keep an eye on what's going on "out there," (or perhaps "in there"), it helps you to relax. Video surveillance of your home or business may mean being able to prevent unauthorized access or being able to monitor otherwise concealed or less-than-convenient sectors of the property.

But security goes far beyond protecting property. With the presence of a video surveillance system, possible intruders can be spotted early on and authorities can be notified. Early warning might well mean the difference in apprehending or at least scaring off intruders *before* they have a chance to enter your home or business, endangering lives or property.

SAFETY

With safety, we are looking at some added protection to hopefully prevent bad things from happening. In the business scenario, this would include monitoring parking lots and other areas susceptible to intruders that employees and others have to enter. Again, the overt presence of video surveillance cameras can really help protect against illegal activity and may help prevent personal violence.

Average criminals do not want their deeds seen on television or recorded on videotape. It just goes against their grain. I don't know, maybe they are self-conscious or have low self-esteem. For whatever the reason, they do tend to be camera shy. Hence, they will not enter areas that have cameras.

Safety is a big part of why people install such observation systems in both homes and places of business. Monitoring children at play, sick or elderly people, and dangerous areas of the home such as poolside and the like, can and will make the home a better place to be and live. For business applications, you can use these systems to monitor harsh environments such as places where hazardous materials, chemicals, or explosives are stored. Don't sell this aspect of video surveillance short.

ILLEGAL OR UNAUTHORIZED ACTIVITY PREVENTION

Both the presence and visibility of video surveillance equipment will go a long way toward making a prospective intruder think twice about hitting your property. Surveillance gear can be worth its proverbial weight in gold when it comes to warding off these not-so-nice guys.

This is like the old adage that says locks are for honest people. That is, crooks know how to get through them, so the locks won't stop them. That is true to a degree, but locks may slow the crooks down, and time is one thing crooks don't have. If those crooks are merely looking for an easy target, they may well pass up your place if you have plenty of locks to go through. The same pertains to the appearance of video surveillance. It is often too risky or just too much trouble, and therefore a great deterrent.

So, in the final analysis, surveillance equipment, like locks, can help prevent break-ins to your property merely by being there. And if an intrusion does happen, you will be alerted and can call police and/or get the perpetrator on videotape.

Thieves are not always intruders, though. For many businesses, particularly retail operations, employees constitute one of the greatest loss threats. Business owners may install surveillance equipment to help prevent employees from "running off with the store." The same principle applies for homeowners whose valuables disappear with the staff. As with the person who is trying to break into your house, any individual with less than honorable intentions will usually think twice before committing illegal deeds in front of a video surveillance system. This deterrence works on another level as well in a business setting. Once someone has been caught with his or her hand in the till, it is a stern reminder to others that "big brother" might be watching.

It is sad that one of the most serious sources of financial loss in many businesses comes from wrongdoing on the part of the employees. Most employees are honest, hardworking, and trustworthy. But it doesn't take too many "bad apples" to put a real dent in the profits. Today, many companies make good use of video surveillance to help prevent this type of loss.

PERFORMANCE MONITORING

On a more positive note, video cameras can also perform a very useful service to employer and employee alike. When placed in strategic locations, the cameras will be able to monitor the day-to-day activities of the business. This will include the performance of the employees, whether it be sales, manufacturing, or some other activity, and help both management and workers do a better job.

This may not sound like a major issue, but it can mean the difference between improved productivity and better wages and layoffs due to poor profits. Everyone can suffer from the bad work habits of a few, but video surveillance can help immensely in correcting those often unintentional problems.

Another advantage of general business surveillance is the opportunity for managers to be in more than one place at a time. Not literally, of course, but being able to keep an eye on the general workday process may be very helpful in spotting potential trouble spots, particularly on the manufacturing or warehouse floor. Is traffic flow jamming up in one area of the warehouse? A video camera might give you advance warning that someone needs to route incoming materiel by less-crowded paths.

This aspect of surveillance is definitely worth keeping in mind. I would strongly encourage management to properly explain the purpose of the video equipment and stress the benefits derived by both parties. Merely throwing up the cameras and telling everyone, "we're watching you" isn't going to endear the program to employees. However, if they understand the career assistance the monitoring can provide, it is far less likely they will feel threatened by the system.

CORPORATE TRAINING AND PROMOTION

This last subject has great merit. It will take equipment above the performance level of the standard surveillance camera and recorder, but you may find any additional expenditure well worth the money. Scenes recorded by prudently placed cameras can provide vignettes of the business's operation that are virtually invaluable in terms of corporate training and/or promotion.

Some of the best corporate publicity tapes I have seen relied heavily on scenes captured during regular company operations. It is hard to stress just how much those scenes added to the overall effectiveness of the production. We have come to live in a very visual world and videotaped "horn tooting" has become almost commonplace.

It takes a little extra pizzazz for your advertisement to get that special attention you desire. Try including shots from your everyday corporate operation. I think you will like the result.

The same arrangement can also furnish priceless footage for training purposes. New hires can be shown tapes of the company in action, and both good practices and bad can be pointed out. Additionally, upper management refresher courses can include scenes of employee activity and company performance. Tapes like these are tremendous tools in evaluating corporate operation and promoting new ideas to improve company efficiency. In the end, you may well do everyone involved a favor through video surveillance.

CONCLUSION

Well, there you have it. I hope this chapter engenders some ideas for you for the application of video surveillance in your business and home. I have presented some thoughts, but there are many other ways the use of video observation gear can be a pragmatic addition to your personal toolbox.

Keep this stuff in mind as it may prove to be the answer to problems that arise down the line. Video's ability to survey and preserve visual images of your property can provide protection, safety, and peace of mind when it comes to the welfare of your loved ones and possessions.

3

Legal Considerations

This chapter discusses some of the possible problems you might encounter when installing and using an electronic surveillance system in your home or business. I will try to cover as many areas as possible; however, there may be other aspects that apply in your area.

This brings me to probably the most important statement I will make regarding this subject: *It is vital to check with your local authorities concerning the legality of **any** type of surveillance system.*

We will discuss some general rules in this chapter, but specific legal guidelines and restrictions are a matter of local jurisdiction. Each town, city, county, and state may dictate what is an acceptable surveillance operation and what is not. In many locations, this will follow fairly generic lines, but there are going to be some areas that impose more stringent regulations. It is always wise to err on the side of caution if you have any doubts regarding your locale. Additionally, rules can (and have been known to) change at anytime, so watch for mention of such changes in your local news media or better yet, check with local authority from time to time to make sure your system still falls in the legal category.

So again, checking with local authorities concerning the legality of surveillance operation is a judicious move. It could well keep you out of trouble with the **law**. Remember the old adage, "ignorance of the law is no excuse."

> **N O T E**
>
> Keep in mind that although you may have captured someone on tape ostensibly committing a crime, the United States practices the principle of "innocent until proven guilty." This doesn't just mean "visible on tape," it means *proven in a court of law*. It may be unwise at any time to publish captured "crimes" (this topic is discussed later in this chapter). But selling or in any way distributing video footage of the alleged crime—for example, for use on a television show of "Crime Stoppers" or "World's Dumbest Criminals," or uploading the footage to the Internet for entertainment purposes—may constitute a violation of the alleged perpetrator's rights. In other words, you may be *sued*. It's as important to know the other person's rights as your own before you decide to make videotapes public.

INCIDENTAL SURVEILLANCE OF PROPERTY NOT YOUR OWN

Unless you are actually invading the privacy of someone else, this one shouldn't present too much of a problem. Here we are concerned with your surveillance cameras catching in their frames property you do not own. For example, you might have a camera watching the front door of your house. If the angle is right, part of your neighbor's house might show in the camera's view.

Is that legal or not? Logically speaking, as long as you are not peering into a window or in some other fashion invading your neighbor's privacy, merely having part of a building in the background of your surveillance shot should not be considered objectionable. However, logic and law are hardly synonymous, thus it would be wise to check the rules.

What happens if your camera, intended to capture or prevent illegal activities on your own grounds, accidentally captures a crime or some other activity on property adjacent to yours? Are you obligated or even allowed to report or share the information? For example, if your front-door camera happens to capture a crime occurring on the street in front of your home, or some illicit activity in your neighbor's driveway, must you surrender the tape to law enforcement authorities? Is the tape legally your property? Can you demand a subpoena before turning it over? Can you sell the tape or upload it to the Internet if you happened to capture some naughty antics of your neighbor, his kids, or his dog? (I'll discuss this topic a bit more shortly.)

Again, remember that controlling authority on these matters is local rule and these regulations may change at any time. It may well be, for example, that due to a previous complaint the city counsel has decided to impose a restriction on not allowing anything but your property to be viewed by your surveillance system. In the way of good old common sense, then it's a good idea to check out the local rules at least once a year.

SURVEYING AND TAPING NON-FAMILY MEMBERS WITH YOUR SYSTEM

This aspect involves the live subjects in your surveillance, and it concerns both the exterior and interior of your property. Is it legal to photograph and video-tape someone at your front door without his or her permission? If an individual enters your home or business, are you obligated to inform that person that he or she is on candid camera? Do the rules apply equally to private property and business locations?

These are some of the questions you should ponder, and if necessary, have answered by the proper agencies in your community. Since we all encounter both hidden and visible TV cameras in many of the stores we frequent, the response to the business side of this question should be that it is acceptable to survey "public access" property.

That may not apply to your area, though, or you may need to obtain a special permit from the city/county/state government(s) to operate an electronic surveillance system. You may also be required to post signs or other visible notices that state you are surveying the property.

Points of concern such as these could cause you problems.

As for your home, the same may be required. It might be necessary to post notice that the home is protected by video surveillance. I mean, we wouldn't want to be unfair with the criminals. You might "entrap" somebody! Laugh you may, but this is an example of just how ridiculous some of these regulations can become.

The problem is that local authorities often take these matters quite seriously and could make an example of you if they feel you were, in any way, disregarding their rules. So be careful and check out the local regulations. Your intention is to protect life and property, not end up in jail.

SURVEYING INDIVIDUALS AROUND A BUSINESS

While we are on this subject, what about the people in and around your business? This would include customers, employees, and others who might be passing by the area. If you protect your parking areas with video cameras, is the privacy of people in those parking areas being violated?

Again, this might seem cut-and-dried. Having cameras watching your parking lot should be considered a security aid to customers, not a privacy violation. However, the local authorities might not see it that way. It will depend greatly on the actions of your local and state lawmakers. For a variety of reasons, they may well have passed bills that create restrictions regarding this aspect of a surveillance system.

As for the interior of the business, it would seem logical that surveillance here would be acceptable. We all see video cameras in the places we shop and understand their necessity. This type of surveillance is frequently designed to prevent shoplifting and, in the long run, helps keep prices lower for us consumers. That should be considered a good thing, but local rule might not agree.

Naturally, there are places you *do not* position video cameras. Restrooms, showers, and dressing rooms are three such areas that come immediately to mind. As for other parts of the business, it may be necessary to give notice that video surveillance is taking place, and covert or hidden cameras may be prohibited.

Additionally, you may have to position all cameras so that they do not include in their view any public access property, such as the street in front of the business, or other establishments across the street. Watching activity outside your business, that doesn't involve business property, may well be taboo.

Don't despair though.... In most localities these are unnecessary concerns. Due to some strange laws and thinking in certain areas, though, it is appropriate that I at least caution you to consider these issues.

As a final note on the problems regarding other property and individuals, for the most part, good old-fashioned "walkin' around sense" will be your best guide. As previously mentioned, you do not want to survey any area where a person's modesty might be infringed upon. I think that should go without saying.

Also, it is not prudent to conduct or continue surveillance that at any point penetrates someone else's property or trespasses on their privacy. Again, this is just plain common sense. If questions do arise about the legality of your system, contact the local authorities.

MAKING VIDEOTAPE PUBLIC

Here we have an area that can really be obscure. Former British Prime Minister Winston Churchill once described the former Soviet Union as a "riddle wrapped in a mystery inside an enigma," and this subject tends to fall into that category.

The rules here are going to be very localized and probably somewhat confusing. Much will depend on what the tape is used for, to whom it will be shown, what is considered "public," for what reason the tape is being made public, and a variety of other concerns. As you can see already, this is going to get very convoluted very quickly.

If the tape is provided to the police as evidence in a burglary, is that considered making the tape public? What if the police decide to release the tape to the local media in an effort to help apprehend the criminal? Is that permissible, does it involve you, and what will be the end result?

While it might seem that you are not responsible for what the police do with the tape once it is turned over to them, that might not be the case. Additionally, in some jurisdictions you may not be required to provide the tape to the police , so what they do with it if you do give it to them may bounce right back into your lap.

I think you get the idea. In this area, be very careful! If necessary, consult an attorney, one who has a great deal of experience in this realm, for advice on how to handle this situation. It may be very difficult to get a definitive decision on this subject from your local city or county offices.

Probably the most serious problem encountered in all this will be the definition or perception of the term "public." Each jurisdiction may well have its own idea of exactly what "making public" means.

COPYING SURVEILLANCE VIDEOTAPES

Much like making your surveillance tapes public, there may well be local restrictions regarding copying of surveillance tapes. Again, restrictions vary so widely that it is quite prudent to check on this if you find a need to copy the tapes you've made.

Any objection to that practice would feasibly hinge on the privacy issue, but there might be other aspects as well. You probably are not going to want to make copies of a vast amount of what is recorded. If you have ever viewed recordings from video surveillance systems, then you know all too well how boring most of this stuff is. It's not generally something to be copied and preserved for posterity.

If you decide to copy business videotapes for training purposes—or you intend for tapes to be used as supporting material in reprimanding or terminating an employee—make sure that you have checked with local authorities and your own company's human resource policies. You don't want to end up getting fired yourself. Even if your company is small, it may (read: should) have written policies regarding the use of employees' images in videotapes. You may need—and probably already have—written permission from each employee appearing in an office training video. If an employee was reprimanded or fired for something captured on tape, you may need written verification that the employee was notified that the unacceptable behavior was captured on tape, *and that the employee has personally viewed such evidence, acknowledged that he or she is the person shown in the tape, and committed the act(s) shown in the tape.*

THE DEFINITE NO-NO'S

This might be another example of stating the obvious, but I feel it is important to mention the following. There are a couple of things that, as far as I know, are illegal in just about all jurisdictions. If in doubt, check locally but don't be insulted if the authorities tend to treat your inquiry rudely.

The first area concerns clandestine video surveillance of someone else's property. That is, hiding a camera on another person's property to observe personal or business activities. I would hope that would not occur to most of my readers, but if it does, **don't do it**! If you get caught, and you probably will, this is (and should be) a guaranteed trip to jail.

A second area involves covert audio monitoring of another's property. In fact, as has been mentioned in this book, audio eavesdropping is always a tricky matter. It is often considered wiretapping even if you are monitoring your own home or business. Hence, it is best to leave out the audio part. Be advised that many video cameras do have audio capability, but I would not advise using that capability, especially in a business environment.

Both of these subjects are clearly invasions of privacy and should be avoided even if you have permission from the party being monitored. That person might want the observation, while others might find it intrusive. All this could lead to a nasty legal tangle that easily could lead to jail time for you. Remember, use common sense and good judgment.

CONCLUSION

Hopefully, this section has familiarized you with some of the possible pitfalls that might be encountered when setting up an electronic surveillance system. As has been said throughout this chapter, if you have any doubts regarding the legality of your equipment, it is best to contact local officials and check on applicable regulations.

For the most part, video surveillance is accepted as a method to help protect your loved ones and property from the "bad guys." If you employ video equipment properly, the chances of running afoul of the law are minimal.

However, there may be statutes in your area that prevent or restrict the use of video cameras for this purpose. A little research into the topic should reveal any limitations.

It has been my experience that in most areas of the country very few, if any, obscure laws exist preventing video surveillance. Where regulations do exist, they should be well known by the authorities.

This section is not intended to scare or dissuade you from setting up a video surveillance system. I merely want to make you aware you that some rules may apply. If you have questions or want to be absolutely sure, consult those who know what rules apply to you and your area.

PART TWO

Audio Surveillance

Let's start with a warning: Because audiotaping may be considered wiretapping in your area, it's crucial that you be fully aware of the laws in your area regarding audio surveillance before you embark on any such project. Once you're over that hurdle, however, you may be pleasantly surprised at some of the fairly sophisticated equipment that's available for this purpose, as discussed in this part of the book.

CHAPTER

4

Audio
Surveillance

I don't think it comes as a big surprise when I say audio surveillance has been around a little longer than video surveillance. That falls into the category of "stating the obvious," but it might surprise some people to know that video surveillance wasn't too far behind audio.

This chapter provides a quick look at how audio devices can be used for surveying your home and business. Since the advent of video, sound monitoring has lost some of its appeal, but under certain circumstances, it can still be useful.

N O T E

As I discussed in Chapter 3, any audio surveillance is potentially problematic, legally speaking. For that reason, this chapter confines itself to a very brief overview of techniques that might be legal in your area. I will not even begin to discuss how to use equipment or techniques that would definitely land you in jail.

MONITORING OR RECORDING SOUND

Probably the first thought that comes to mind, regarding audio surveillance, is the proverbial "bugging." Unfortunately, this area of sound surveillance has seen an increase in recent years, as opposed to a drop-off. In most instances, though, it is illegal! I would not recommend it even on your own property. Some states see any type of audio observation as wiretapping, and often it can carry a stiff penalty that involves large fines and/or jail time. If you plan for any reason to install clandestine microphones, hard-wired or wireless, I would strongly advise you to either check with local authorities regarding the law or consult an attorney knowledgeable in this area.

However, small tape recorders, microphones, or other sound-detecting devices in plain sight are a different story, especially if you inform people entering the effective area of the devices. Now you are not monitoring or recording audio in a covert fashion, and those caught on the tape or surveillance system have little legal recourse.

With the arrival of the condenser/FET elements, microphones have become very small in the last couple of decades. They are not only small, but also very sensitive compared to some of their older dynamic and crystal cousins. This is made to order for both bugs and hidden hard-wired setups. It also allows the microphone to be placed directly in a transmitter package for a far more compact and pragmatic listening device.

When it comes to this application, wireless transmitters come in a variety of shapes sizes and flavors and you will be able to find something that will do the job. Some of these transmitters work in the normal AM and FM broadcast bands, some are in bands shared with other radio services and come with a dedicated receiver, and others operate in the commercial business bands. Most are relatively low power, which can be an asset, but some will send a signal quite a distance.

Some transmitters will require a license to operate, while many fall within the "low power" regulations set by the Federal Communications Commission (FCC) and do not require licensing. In most situations, lower power is probably going to be the best way to go, as often you do not want anyone else listening to what the microphone is picking up. Also, in a covert scenario, stronger power lends itself to being more easily detected by the very people you are listening to and don't want aware of the device.

Thanks to semiconductors, and their low-power requirements, much of this stuff can be operated on batteries. That, of course, makes them very portable and convenient to use. Unlike the hard-wired arrangement, a small container the size of a cigarette package can house the microphone, transmitter, and battery and be easily placed and removed. This characteristic allows for far more flexibility. You will, however, be limited by the capacity of the battery.

Additionally, the cost of such equipment has dropped to the point where the loss of discovered devices is almost negligible. Some of these transmitters have practically become disposable. In fact, you may not want to even try and recover the listening devices. Be reminded, though, much of this activity falls into areas of illegal operation.

TELEPHONE AND OTHER OVERT APPLICATIONS

Some of the more overt applications of listening devices utilize the telephone system. There are several devices on the market that allow you to phone your own home and "listen" to activity in various rooms of the house. Here, audio surveillance can help monitor elderly or sick individuals, children, and pets, or just permit you to see if all sounds well at home.

Another telephone gadget allows you to record both sides of any conversation you have on your telephone. This usually works on any phone connected to the system and as long as one of the parties involved is aware of the "tap," it's normally legal.

I have spoken with authorities at the federal level, and asked why this differs from bugging your own office and have yet to get a really acceptable answer. For whatever reason(s), generally speaking, bugging your own telephone seems to be exempt from wiretapping laws. Again, be on the safe side and research your state and local community. Check out Chapter 3 for more information on legal issues.

On a last note, the Internet has opened another avenue of opportunity. There are systems that will permit you to monitor your home, both with audio and video, virtually anywhere in the world. That is, anywhere you can get a telephone, satellite, or radio signal, and these days, that is just about everywhere.

These systems again allow a person to call home and monitor any activity occurring in the home.

CONCLUSION

Audio surveillance can be, in some situations, very beneficial. I personally feel you will find greater versatility in video systems, but there may well be times when an audio "ear" might do the trick. The cost will be substantially less than video equipment, and if all you need is sound, then this approach makes sense.

At the risk of sounding redundant, or being a pest, I will again caution you to check local regulations regarding audio surveillance. The statutes for each town, city, or borough can vary so much that relying on state and/or federal laws may be the wrong way to go (see Chapter 3 for a discussion of legal issues involved in audio surveillance and wiretapping).

PART THREE

Video Surveillance

Because the majority of surveillance activity involves seeing a picture of what's happening, a discussion of the selection of video devices and accessories is a must. With the arrival of "miniaturization," cameras can be hidden in pens, eyeglasses, videotapes, wall paintings- and that doesn't even begin to cover the add-ons and other accessories. The chapters in this section explain what's available and hopefully can help you to determine what you'll need for your particular application.

Additionally, I'll talk about the numerous and ingenious ways to get the picture to where you want it, including both hard-wired and wireless methods. So, sit back and enjoy these chapters, as they'll enlighten you to all the neat things you can do with video surveillance.

5

Video Surveillance Overview

As I mentioned in Chapter 1, a friendly war (is that an oxymoron?) was waged during the early years of television's development between the armies of "electronic TV" and "mechanical TV." The boys with the spinning disks fought a good fight, but in the end electronic scanning won out. This was largely due to the reliability and better image quality afforded by the latter method.

Anyway, when the smoked cleared and the surrender was signed, what we know and love as television was left the victor. Okay, enough of the battle analogies. Let's move on to the way in which electronic television works.

VIDEO FUNDAMENTALS

Probably the best place to start is with the video concept itself. That is, what makes up one line of a color video signal. First off, let me say that electronic video is complex. There is no way to get around that, but that fact shouldn't scare you. With today's technology and components, the complexity is easily managed.

Figure 5.1 shows a single line of National Television Standards Committee (NTSC) video. Notice that this is a horizontal scan line. It begins with 1.54 microseconds of blank space (no signal) known as the *front porch*. It has been said

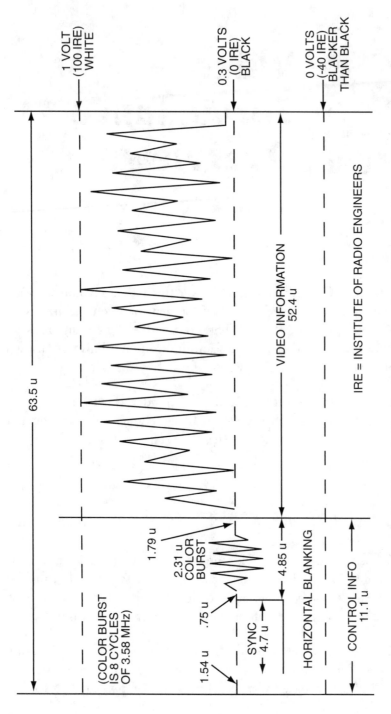

FIGURE 5.1 One horizontal scan line (NTSC color).

that the three most important elements of video are timing, timing, and, most of all, timing. As we make our way through this narration, I think you will begin to understand why that is true.

The next segment of the line is the synchronization signals and they last exactly 4.7 microseconds. This is followed by a 0.75 microsecond blank space, commonly known as the breeze-way, and the 2.31 microsecond color subcarrier, or *color burst*, signal. Here, a break with video timing tenacity is found, as the color burst can be eight to eleven cycles of 3.579545MHz +/– 10 Hz. These departures from the very stringent video timing scheme are few and far between.

The color burst signal is then followed by another blank period 1.79 microseconds in length, known as the back porch. This completes the first 11.1 microseconds of the NTSC line, which contains all of the color and synchronization information that keeps the final image pure.

Any variance in this routine is going to result in an imperfect picture on the television screen. If the sync timing is off, the image will be unintelligible. If the color information is not on target, the people in the picture may well have green skin and purple hair. (You may well know someone with green skin and purple hair. I have seen a few like that myself. But that is not the norm.)

The bottom line here is that the information in the signal is far less important than the fact that it arrives where it is supposed to be **on time**! It follows that the information should also depart **on time** and keep all the timing in line. With that understood, video becomes a much easier subject to comprehend and work with.

Back to Figure 5.1. Now that we have the synchronization and control signals out of the way, the remaining 52.4 microseconds of this single horizontal line are devoted to primarily video information. That is, the picture or image, which is black-and-white in nature. It is also an amplitude modulated (AM) signal that runs between 0.3 and 1 volt peak to peak.

A 1 volt signal measurement represents the *white* or *white balance* level, while 0.3 volts is *black*. Below 0.3 volts is an area known as *blacker than black*, but we will get to that in a minute. Naturally, everything in between those two benchmarks represents the *gray scale* of the image—the shades of gray that make up the picture. Remember, the video information is in a black-and-white format.

Now, I said "primarily video information" because the last 1.79 microseconds of the signal is another blank space known as the *back porch*. This is inserted at the end to tell your television set that the horizontal line is completed. That, in turn, triggers the system to move to the next horizontal line.

Also, at the end of a full screen scan, the back porch initiates vertical retrace, taking the CRT beam back to the top of the screen for the beginning of the next field of video information. Video is formatted in *frames* that occur every 1/30 of a second, and each frame consists of two *fields* 1/60 of a second in duration. More will be said about this in the monitor section of this chapter.

In Figure 5.2, a typical "broadcast channel" is illustrated. Each of the channels a standard television can receive is 6 megahertz wide. This is a lot of frequency spectrum, especially since we are talking about 81 channels total, but it is necessary for the analog arrangement contrived by the NTSC.

Figure 5.2 deals with broadcast channel 3 (60 to 66 megahertz), so let's talk about how this 6 megahertz is divided up. The first 2.5 megahertz are devoted to the video image, or *luminance* signal. You will hear the term *center video* used in conjunction with the NTSC format, and that refers to the center frequency of the luminance segment. In this case, that is 61.25 megahertz, or 60 megahertz plus one half of the 2.5-megahertz segment.

While the entire 2.5 megahertz is used to convey the black-and-white picture information, the center video frequency establishes an *AM luminance subcarrier*. This, of course, assists the television receiver in isolating what information is coming in. It also helps establish the luminance portion of the 6-megahertz signal and is the frequency you will want to tune a video transmitter to if you plan on using broadcast channel 3 as your transmit frequency.

The next section of this channel is used for both synchronization and control signals and color (*chrominance*) input. This may look confusing in the diagram, as it might appear that this data is being sent at the same time. In reality, a careful phasing arrangement is employed to interlace the two types of information. In the same fashion as multiplexing digital displays so that only one display segment is actually on at any given time, only the chrominance or sync information is being transmitted at any given time. Here again, timing, timing, and most of all timing!

THIS IS AN OUTLINE OF VHF CHANNEL 3 AT 60 TO 66 MHz

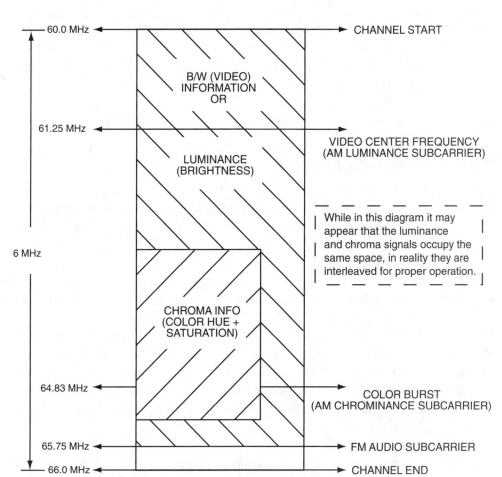

FIGURE 5.2 Typical 6 MHz broadcast channel (NTSC).

The color burst, or *chrominance subcarrier*, is going to occur at 4.83 megahertz above the lower channel edge, or in this case, at 64.83 megahertz. Like the center video subcarrier, color burst helps establish an information area and in this case, keeps the color orientation in line.

The last segment of a 6-megahertz video channel is the audio information. What would television be without the sound? This is *frequency modulation* (FM) and occupies the last 0.25 megahertz of the channel. In our example, the audio subcarrier is found at 65.75 megahertz, or 4.5 megahertz above center video. As Figure 5.2 shows, after the sound signal the channel ends.

So there you have explanations of both a single line of horizontal scan, and how the 6-megahertz channels are segmented.

TRANSMISSION LINKS

Let me now introduce you to the methods used to relay the information from the television camera to the TV receiver. There are two ways to go about this: closed-circuit TV (CCTV), which employs cables, or *hard wires*, to carry the camera signal to the monitor, and wireless, where a radio frequency transmitter is used to send the camera data out into the air. The data is then "snatched" from the air by a television receiver designed to detect the necessary radio frequency signals.

Both methods have their merits, but the important aspect of all this is that both do nothing more than relay the vital luminance, chrominance, sync, and audio signals generated by the television camera and translated by the receiver. Thus, even the more complex wireless approach is not nearly as intricate as the camera and receiver. This makes things a lot easier in terms of getting the picture and sound to their destination.

Naturally, closed-circuit systems are merely a matter of running cable from the camera to the monitor. Unless you have extremely long runs, which requires cable amplification, closed circuit is definitely not rocket science.

Radio transmitters, on the other hand, present a little more of a challenge. You will have to be able to radiate an RF signal into the atmosphere, and it will have to be modulated with both the video and audio information. That is presuming you do want to transmit both video and audio. I say that as I have built several transmitters that send only the video. When audio is not needed, which is often the case with surveillance arrangements, the video-only transmitters are much easier to design and build. Transmitters are the missing link between the camera and monitor.

RECEIVERS AND MONITORS

What we are interested in here is both television receivers and dedicated video monitors. As might be expected, they are very similar devices with rather subtle distinctions. These distinctions will be covered in greater depth later on.

At the outset, let me say that the receivers and monitors function on the same principle as the cameras, in terms of sync, video, audio, and so forth. Timing, timing, timing! They just do it in reverse. That is, instead of generating the various signals, the receivers decode them, and it is that decoded information that allows the monitors to reproduce an accurate version of the original subject.

> **N O T E**
>
> I will use the terms *receiver* and *monitor* interchangeably at times to make reading this stuff a little easier on you (a little less repetitive hopefully). Fear not, a clear distinction will be made between the two when necessary.

The primary difference in the two systems is that TV receivers (television sets) incorporate a tuning section that, like any radio, detects electromagnetic waves at the frequency it is tuned to. However, transmitted TV signals are *composite* in nature. What this means is that all the video, control, sync, and audio information is transferred as one signal, which the tuner receives, separates into the component parts, and then sends to the appropriate monitor sections. For example, the video component is routed to the video amplification circuit, while audio goes to the audio amplifier. This was, in all reality, the only way the NTSC could have dealt with the task of developing a standard for television when television itself was in its infancy, and that standard has served us well for over fifty years. But the video output of the camera, which separates the control, sync, and video signals from the audio signals, cannot be directly connected to a standard, or antenna input, television receiver. The tuner doesn't know what to do when with the already-split signal.

Dedicated, closed-circuit monitors, on the other hand, have separate video and audio inputs that correspond to the individual signals from the camera design.

Each component is fed directly to the segment of the monitor circuit that handles that information. In a closed-circuit scenario, the individual inputs work quite well and often produce a superior picture.

The answer to the dilemma of how to watch what your camera captures on your standard television lies in generating an RF signal the TV will recognize. The device that does this is often called a *video modulator* and, by the way, is the subject of one project in Chapter 13.

However it gets there, when the signal arrives at the TV or monitor, the set looks for the relevant control data. When the front porch is detected, the monitor knows to start a horizontal video line. From there, the various sync and regulation signals keep the horizontal line on track as it makes its way through the process.

Most monitors used for surveillance and all standard NTSC-format television receivers employ a scanning scheme known as *interlacing*. In this scenario, half the screen (either all odd-numbered lines or all even-numbered lines) is scanned, or *painted*, by the electron beam on each pass. This is known as one *field*. The reason for this approach involves the visual persistence of the phosphorescent material used by the CRT. For a more detailed explanation of interlacing, see Chapter 8.

At the time the NTSC was setting forth the standards for commercial television, had they tried to scan the entire screen in one pass, the phosphorescent chemicals would not have the visual staying power to remain illuminated until the next pass refreshed them. Hence, the standards commission settled for the "two fields equal one frame" arrangement. In this fashion, the refresh signal arrives twice as fast and easily repaints the screen before it fades from the first pass.

However, this caused some other problems. In order for it to work, the television has to be able to distinguish between field one and field two in order to keep the video information straight and the picture correct. This was solved by the fact that a full scan is 525 lines. Thus, a field is $262^1/_2$ lines. Those half lines possessed the answer to the quandary. Field one starts with a full line and ends with a half line, while field two does the opposite. Now the monitor knows which field it is dealing with and where to put the contained data.

Following the front porch, the television sees the sync signals that set up the timing. Do not ever forget timing when dealing with video. Next, if color information is present, the color burst will establish the pattern and arrangement. Finally, the actual video information is presented as an amplitude modulated signal that tells the monitor what the scene looks like. When all this is put together, you have your television picture.

Whether the data is coming directly into a dedicated A/V port or is the product of a dissected composite signal, ultimately, the procedure for displaying the final image is going to be the same. With all the elements in place, the process works great.

CONCLUSION

Hopefully, this narration has given you some idea of how video works. It should also have impressed upon you the utmost importance proper timing has in the video process. This is, of course, a quick and dirty version, and there is substantially more detail if you want to pursue it. If not, the proceeding should leave you with at least a decent understanding of what is going on inside your TV set.

6

Video Equipment: Basic

At its most basic level, video surveillance consists of two pieces of equipment: the camera that captures the images, and the monitor that you use to see those images. It doesn't really have to be more complicated than that. You don't even need to record the images captured by the camera—the stand-alone equipment we are concerned with here is a little different from the run-of-the-mill camcorder you might buy at Target. Traditional camcorders like you might use for home movies usually have Audio/Video outputs and could be used in a surveillance system, but that is an expensive way to go. This book primarily deals with cameras that send their video signal to a separate VCR, monitor, or radio frequency transmitter, or all of the above. The individual camera units are far cheaper, smaller, and more practical for home/business surveillance. If you want to record the image captured by one of these cameras, you'll need to add a VCR or other piece of recording equipment to your system. This chapter introduces the most basic pieces of equipment, the camera and monitor, and discusses some of the issues involved in choosing the right equipment for your particular situation.

CAMERAS

Whenever the subject of video cameras comes up, I often relate the tale of the first TV camera I ever built. Now, I'm talking many, many moons ago. It was a Vidicon-based black-and-white job that used all electron tubes. Other than a handful of rectifiers, there wasn't a semiconductor in the unit. Once completed, it was hailed as a marvel of its time, because the overall dimensions were about ten inches by eight inches by six inches. I was frequently asked how I got such a complicated project into such a small container.

Of course, solid-state cameras were just around the corner, and they would put the size of my dinosaur to shame. But for its time, it was impressive. What was not impressive was the amount of time that it took to build and align the thing. I spent over a year on the project, and only a few months were involved in actually putting the camera together. I spent the rest of the time trying to get the picture to stop rolling or just trying to get a picture at all. One of the problems was the vacuum tubes. Those little wonders get very hot when power is applied, and that heat continually changes their characteristics.

To say that video electronics is temperamental is being kind. The timing parameters of any video circuit are so critical that even small changes in the operating attributes of a component can wreak havoc with the performance of the circuit. Once I did get the camera working, I was forever readjusting it to keep it working properly.

Naturally, the arrival of semiconductors into the picture (excuse the pun) eliminated many of these difficulties. Semiconductors created some stability in the functionality of the circuit components, and that made all the difference. Further, the invention and perfection of the charge-coupled device (CCD), or solid-state imager, brought the video camera into the modern world (see Figure 6.1 and Figure 6.2).

With that rather long story out of the way, let me get down to the heart of this section—what is available today.

CCD Versus CMOS

As mentioned in Chapter 5, there are two primary types of imaging devices used today: the charge-coupled device (CCD) and the Complimentary Metal Oxide Semiconductor device, better known as the CMOS imager. As of this writing,

PHOTO COURTESY OF SUPERCIRCUITS, INC. USED BY PERMISSION.

FIGURE 6.1 A collection of various color and B&W television cameras. The smaller board cameras are in the front, with an assortment of encased units toward the back.

PHOTO COURTESY OF SUPERCIRCUITS, INC. USED BY PERMISSION.

FIGURE 6.2 The older-style Vidicon tube sensor (above) versus a more modern CCD imager (below).

CCDs rule due to their better resolution and color saturation, but CMOS is not far behind. Improvements in this technology have been nothing short of phenomenal, and it won't surprise me to see these imagers take the lead, perhaps in the near future.

In the way of a quick and dirty comparison, CCD cameras produce better pictures in terms of clarity, light sensitivity and, in the case of the color versions, color saturation, than the present-day CMOS units. However, CCD does require considerably more power. A CMOS camera will usually run on ten to fifty milliamperes, while its CCD cousin will demand 100 to 150 milliamps. That makes CMOS very desirable for portable or battery operation.

Also, at this time, the CMOS cameras tend to be cheaper than CCDs. The primary reason for the reduced cost is that, if you look at one of these board cameras, which are open circuit and require some sort of enclosure, you will see that all the electronics (image detector, sync, and amplification) are contained in a single integrated circuit (IC), or chip. There will only be a handful of support components, such as a crystal and capacitors, soldered to the rest of the printed circuit board (PCB). In this fashion, the expense can be greatly reduced. You can see there are pluses and minuses to each technology.

Camera Resolution

The resolution, or number of horizontal lines the camera produces, is always a subject of interest when these micro video cameras are discussed. For the record, the analog television receiver you have in your home has a resolution of 525 lines. That is to say, the picture you view is divided into 525 individual horizontal scan lines. Take a magnifying glass to the screen, and you will see what I mean.

However, this doesn't mean a TV camera has to produce the same resolution. With most of the board cameras, the resolution is in the 330 to 410-line range. Naturally, the greater the number of lines, the clearer and sharper the picture is going to be. As the technology gets better, the resolution of these small wonders is improving, but for the most part, the middle-range monochrome devices have 380 to 410 lines of resolution, while the color cameras range from 330 to 380 lines (see Figure 6.3).

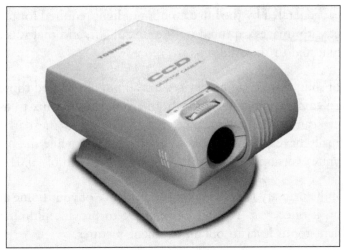

PHOTO COURTESY OF SUPERCIRCUITS, INC. USED BY PERMISSION.

FIGURE 6.3 An encased CCD color video camera sits nicely on a desk.

There are exceptions (some of which are discussed in Chapter 8), but for many applications the 380-line clarity is remarkably good. With resolution less than the 525-line standard, the camera picture information is spread over more than one TV horizontal line, thus making it a little less distinct. Normally, unless you view the screen at a very close distance, it is hard to distinguish this lower resolution. You might expect the rule to be, the higher the resolution, the higher the camera price, but even that is changing as new models hit the market.

Lux Rating

Another important factor in video camera specifications is the lux rating. This is a measurement of the camera's sensitivity to light. One *lux* is defined as the amount of light produced by one candle at a distance of one foot. The lower the lux rating, the better the camera sees in low light settings.

The older image tubes (iconoscopes, orthicons, vidicons, etc.) had very poor sensitivity to light. This did improve as newer tubes arrived, but there are horror stories regarding film sets catching fire and makeup melting off the faces of actors due to the heat generated by the huge amount of light required for taping. These tubes did best with images captured in bright sunlight, and that was about what had to be replicated in the studio.

The advent of solid-state imaging devices has all but eliminated those problems. Today's board cameras range in rating from around twenty lux to as little 0.01 lux. The latter can, without assistance, practically see in the dark. The color cameras normally have ratings about twenty to five lux, while the monochrome (black-and-white) versions fall into the five to that incredible 0.01 category.

This is one of the primary factors making surveillance of your home and business far easier than it once was. You no longer have to provide illumination above normal ambient room level to obtain excellent pictures.

Color Balance

On a last note regarding camera performance, let me mention color balance. For those who have a background in photography, this is a familiar subject. The term *color balance* in this context refers to the ability of the camera to see objects in their proper color. Light is not always the same from source to source, and different sources produce light of different temperatures. Here, temperature encompasses the Kelvin rating of the light. For example, the sun is around 7,000 degrees Kelvin, while tungsten light (a light bulb) is about 3,200 degrees Kelvin. This difference in temperature makes an enormous difference in the shade of objects exposed to the light. The normal or pure colors we see when an object is in sunlight will become warmer, moving more toward red or orange, under a tungsten source.

It is the job of the camera to adjust for various types of light. With the older tube imagers, this generally had to be done with filters placed in front of the lens. Modern semiconductor imagers accomplish this through a digital signal process-ing (DSP) scheme known as *white balance*. With most of the CCD and CMOS cameras, white balance is an automatic function of the circuit. It makes certain that the picture we see on the monitor appears to be correct in color regardless of the lighting source.

PHOTO COURTESY OF SUPERCIRCUITS, INC. USED BY PERMISSION.

FIGURE 6.4 A very small CCD color board camera. This one has a lux rating of less than five and 380 lines of horizontal resolution.

Again, this color balancing ability of the solid-state cameras tends to make things easier. No more gel filters to correct for lighting imbalances. Virtually any type of white light can be used to illuminate the scene and subject.

The sizes of these cameras are nothing short of extraordinary. Remember the description of my first tube camera? I now have one color CCD device that is the size of a postage stamp (a little deeper, of course). It is less than an inch square! (See Figure 6.4). Another black-and-white CMOS device is roughly the size of a sugar cube, and that includes all the electronics necessary.

These cameras are also sensitive to infrared light, which allows them to use infrared light emitting diodes (LEDs) as a source of illumination. Often the diodes are installed in a circle around the lens and this permits the camera to see in the dark. Aim your remote control at one of them and see what happens!

Pardon my enthusiasm! I guess you would have to have gone through my experience with the old vidicon cameras to truly appreciate the new solid-state equipment. I have been working with solid state for at least ten years, and I still can't help but be astounded by the little gems. It seems that on an almost daily basis newer and more amazing cameras appear on the market (see Figure 6.5).

PHOTO COURTESY OF SUPERCIRCUITS, INC. USED BY PERMISSION.

FIGURE 6.5   Compare the older vidicon black-and-white camera
(left) to the newer CCD B&W board camera (right).

Cases for Board Cameras

The devices I have been discussing are called board cameras, as they are open
circuit and do require some sort of enclosure to protect them from both physical
and static electric damage.

In addition to the open configuration, many of the distributors offer these board
cameras already in cases. These range from very small plastic or metal boxes that
just hold the board, to larger, more visible enclosures that are helpful in overt
surveillance applications. There are times when you want people to be aware of
the presence of these cameras, as I discussed in earlier chapters. (See Figure 6.6
and Figure 6.7.)

These cases usually assume the stereotypical elongated box shape, or a slightly
rounded shape, depending on the manufacturer. They are normally in the range
of two inches square by four to five inches long , which makes them ideal for
overt surveillance of a business or home.

PHOTO COURTESY OF SUPERCIRCUITS, INC. USED BY PERMISSION.

FIGURE 6.6 A small color surveillance camera mounted in a protective case. Note the ball arrangement at the bottom that's used to mount the camera.

PHOTO COURTESY OF SUPERCIRCUITS, INC. USED BY PERMISSION.

FIGURE 6.7 A larger camera case. Potential rule breakers will know they are being watched.

Another popular shape is the round lipstick or bullet camera. These are usually weatherproof, which means they will withstand some moisture and can be used outdoors if you are careful where you place them (it is not a good ideas to expose them to frequent torrential rainstorms). See Chapter 12 for details on protecting and maintaining your equipment. Their name comes from the fact that if you

PHOTO COURTESY OF SUPERCIRCUITS, INC. USED BY PERMISSION.

FIGURE 6.8   Lipstick or bullet camera.

use your imagination, they do look a little like a tube of lipstick or a bullet. (See Figure 6.8)

However, these are just a few of the many configurations you will find in solid-state cameras. One other aspect that may be a consideration is the camera lens, but which lens will work best for your situation is much easier to demonstrate than to discuss. When you start shopping for a camera, ask your distributor to literally show you what different lenses see. This will give you a much better understanding of what different lenses will be able to do for you than any discussion I could offer. I will get into the more exotic stuff down the line. For now, this gives you a reasonable image of the most common devices available.

MONITORS

Naturally, a video camera isn't of much use unless you have some way of seeing what it is seeing, which brings us to our next subject: monitoring devices.

TV Monitors

While semiconductor systems are gaining strength all the time, the good old cathode-ray tube (CRT) still dominates the scene when it comes to monitoring video images. Of course, I'm talking about the television set. Virtually all analog cameras made today adhere to the National Television Standards Committee (NTSC) format of 525 horizontal lines and a composite input signal (that is, all the information regarding the picture, sync, luminance, chrominance, etc., phased onto a single input line). Not many of the cameras actually produce 525-line resolution, but that was discussed earlier, in the Camera Resolution section.

With that in mind, let's look at what is available in the way of monitors. Probably the mental image you get when the term monitor is mentioned is a rectangular box with a large picture area on the front. That is about what most of these devices look like. Some resemble computer monitors, while others have a more traditional TV style.

In the end, they all accomplish the same goal, and that is to display the signal (picture) from a video camera or other NTSC video device, such as a videocassette recorder (VCR). The primary difference between monitors is in the size of the picture area. At present, CRTs range from around five inches diagonally to twenty-seven inches and larger. For surveillance purposes, the nine, ten, twelve, and nineteen-inch tubes seem to be the most functional, depending on how many cameras are being monitored, how much space you have in your monitoring station, and other such issues.

Now, there is a difference between the television set you use to watch local and network programming, and a dedicated video monitor. The TV has a tuner built in that allows you to receive stations that are transmitting a signal (this includes cable systems). The dedicated monitor does not have the tuner section, and so its input has to be a standard audio/video (A/V) NTSC signal. I know this can get a little confusing, as all television signals are NTSC formatted, but when you are monitoring the signal directly from a video camera, you do not want that signal to encounter a tuning section. The camera does not generate a radio frequency (RF) signal, which can be tuned, merely a composite output.

There is an accessory known as the *video modulator* that will take the composite signal and ride it on RF. This will allow you to view the images your camera captures via a standard television set. You can learn lots more about this—I've included a video modulator in the *Projects* section of this book.

The video modulator is the answer for using regular TV sets with video cameras. Many individuals run the camera through a VCR and then to the TV and that works great, especially if you want to record what the camera sees. The reason it works is that the VCR has a built-in modulator. However, you should be aware that standard televisions generally do not have the higher resolution, or picture quality, of dedicated monitors. Normally you will get a far better picture with video monitors. I know I'm using a lot of qualifiers here, but there are always exceptions to the rule. For the most part, the standards will apply to your situation.

Naturally, both TVs and dedicated video monitors come in both monochrome and color versions. Of course, unless you are looking at a five-inch model, standard monochrome televisions are getting scarcer all the time. Anyway, color monitors permit you to view the images either way. It goes without saying that you have to have color cameras to get color images, doesn't it?

Liquid Crystal Display (LCD) Monitors

Now comes the newest addition to the monitor scene—the Liquid Crystal Display, or LCD. These are compact, energy-efficient, semiconductor displays that have significant advantages. Their size enables them to be used in areas that simply do not allow for a CRT device. This is due to the fact that LCD monitors are solid state, thus fairly flat. They do not have the obtrusive electron-gun neck that conventional picture tubes possess.

Additionally, the power requirements of LCD monitors are minimal compared to the CRT systems. Most LCD displays run off twelve volts direct current (DC), while some do need a second voltage (usually around eighteen VDC) for the fluorescent backlight assembly. Also, without the vacuum tube, they do not experience the heat problems associated with the CRT.

However, nothing is perfect. To date, the resolution of most LCD units is less than that of a cathode-ray tube. It is getting better, though, as the technology improves.

Another problem with LCD displays is viewing angle. The picture on an LCD monitor will virtually vanish if the viewing angle exceeds certain parameters. That means that you have to look at the screen straight on generally. If you get too high or low, or off too far to either side, you will lose the picture.

All of these obstacles are being addressed by research and development (R&D) folks, and a great deal of progress has been achieved in just the last decade. I remember well the first black-and-white LCDs that produced essentially unintelligible pictures. I also remember the first color LCD screens that displayed what appeared to be a collage of small colored dots rather than an image. Today, the quality of LCD has improved, through active matrix design, to the point where both the surveillance market and the computer industry can make good use of them.

Computer Monitors

Speaking of computers, they can also be used to monitor video cameras. This does require additional hardware and software, but it can provide a very handy way to check your surveillance system. The hardware usually consists of a special card (peripheral computer board) that fits into an open slot and will often produce both composite and modulated output signals. That allows for monitoring on either, or both, a standard TV and a dedicated display (see Figure 6.9).

In addition to these video capture cards, there is another device that has between four and sixteen video inputs viewable through a dial-up modem or Ethernet network. With this technology, it is easy to export video onto the viewing computer for saving, capturing, or simply spot-checking your facility remotely. With the advent of the Internet it is also possible to put live video directly to the Web. You may have heard of child-care centers where the parents can log onto a secure Web site to see what their kids are doing at that exact moment in time. If you need to be able to check in anytime, from anywhere in the world (well, pretty much), this is an option to keep in mind.

Frequently, the card has a built-in tuner as well. This permits the user to watch local or cable television programming when the card is connected to the appropriate input (antenna or cable hookup). In short, this strategy affords considerable versatility regarding your computer and surveillance system.

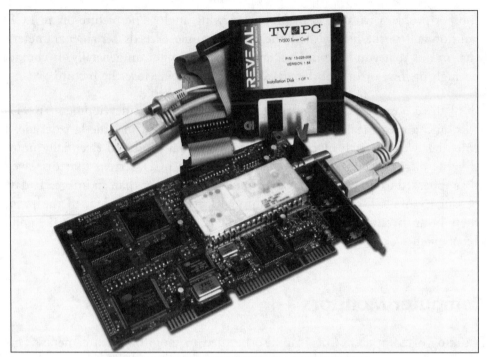

FIGURE 6.9 A typical video/television computer card used to capture video images and receive broadcast TV signals. It comes complete with cables and software disk.

CONCLUSION

Hopefully, this chapter has familiarized you with the video surveillance components you will need and be working with, at least as far as a basic camera to monitor the system is concerned.

Don't let this intimidate you, as none of it is rocket science. Once you get ready to purchase and install the equipment, what we have covered here will all fall into place. Most cameras come with wiring harnesses that allow for very easy plug and play hookup, and the monitors, VCRs, and other devices are well marked to make installation simple and convenient.

Video Equipment: Intermediate

The basic camera-plus-monitor setup discussed in Chapter 6 will accommodate most home uses, but your business or property may need more ingenious devices. This chapter explores some of this more interesting step-up gadgetry. Obviously, the equipment discussed here will require a larger investment of money, time, effort, and knowledge than the simple setup, but it should also provide more security.

MECHANICAL AND ELECTRONIC TRIP DEVICES

This section briefly covers some of the accessory items that can be used to turn the surveillance on or off. They can be home brew or purchased, and mechanical or electronic in nature. All have the same purpose and that is to activate the equipment you use to survey your home or business.

Mechanical trips tend to be somewhat Rube Goldberg in design and less than reliable. These include such things as door brackets that activate switches, both the mechanical switches that are activated when a door is opened or closed, and the magnetic units that cause a reed-style switch to close when a magnet is brought close to it. Both types are readily available from electronics mail order companies and stores. There are levers and even wires stretched across a room that can do the same thing. In the case of the latter, the term *trip* could well become literal.

Most of this stuff is not dependable and hardly worth the effort, especially in light of how inexpensive and easy to use most electronic devices are. If you are good at building electronic equipment or kits, you will be able to make much of the gear you will need. Even if you are not all that handy, the cost of factory-assembled equipment is well within most budgets.

NOTE

In the way of a construction project, an automatic tape recorder switch for telephone lines has been included in Chapter 15. This fits into the category of trip devices, but involves audio surveillance, as opposed to video.

In the way of electronic gadgets that activate surveillance systems, they can be motion activated, telephone activated, computer activated, sound and light activated, alarm-system activated, or time activated.

Motion Activation

Motion activation is probably the most commonly employed method in today's surveillance. It most often uses the same infrared detection device (passive infrared, or PIR) that turns on the floodlights in your yard or controls the porch light over your front door. The heat of an approaching body triggers these units, and they have proven quite reliable within about a fifteen to twenty-five-foot range, depending on the unit you purchase. Another method is to use a low-power RADAR system. Usually Doppler in nature, the RADAR detects any approaching object and triggers the surveillance gear.

A note of caution: Motion detectors are not unreliable, but it is possible to side-step a motion-activated switch. It is not easy, but if someone knows what he or she is doing, it can be done.

All in all, for a cheap means of observing either interior or exterior movement, the infrared systems are hard to beat. You may end up with a lot of videotape of the neighbor's dog, but then again, you might be able to provide the police with a picture of the burglar who tried to break into your business. That last scenario could make this type of detection well worth the money.

Another kind of motion detection switch is digital video motion detection. You have to have a computer hooked up to your camera, which basically samples the video coming in. If a certain amount of pixels change in the field of view, a set of "dry" contacts is generated to then in turn activate whatever device or devices you have connected to the switch. A nice feature of these devices is that you can usually mask off areas in the view to prevent false activation. For example, if your camera is watching a room with a fish tank, you can mask the fish tank out, basically telling the computer to ignore it, to prevent false start-ups. One application this system is not well suited for is outdoors. In uncontrollable environments you obviously get more erroneous activations than you would have indoors. But if you set it up properly in a controlled environment, this is a very effective system.

Telephone Activation

Telephone-activated systems are similar in design to motion detectors, except they utilize your telephone system and lines rather than motion. That is, when you want to initiate surveillance, you simply phone your home or business and tell the system to "get with it," normally by entering a code using your touch-tone phone. This system allows you to start surveillance anytime you want to, as opposed to a timer arrangement. You wouldn't use this system to detect break-ins, obviously, but it is very handy for periodic surveillance of your business, to check for irregularities, look in on employees, and so on. Simple in nature, highly reliable, and hard to defeat, since a person would have to actually get into the phone system to disable the surveillance, this form of protection assuredly has its purpose in certain observation applications. The big drawback to telephone activation is cost. While prices are declining, by comparison, you will spend considerably more for this type of arrangement.

Computer Activation

Computer activated is the next method, and it is similar to the other two in the respect that it powers up your surveillance apparatus by means of an external stimulus. In this method, you control the system by means of your personal computer. A start-up program permits you to schedule activation however you need it, and most allow for great flexibility concerning activation/ deactivation times. Most of these start-up programs do allow for external detection devices to be connected to the computer so your system can be both software and detector driven.

With computer monitoring, additional software and hardware are necessary, and that will, of course, add to the cost. As always, the degree of security you desire may justify this expense.

Light or Sound Activation

Another simple and cheap type of activation can be achieved through changes in ambient light or sound. Light activation works when someone or something passes through a beam of light, breaking the light, and putting your whole operation into motion. You see these systems all the time in heist movies, but don't believe everything you see. These systems are not easy to sneak through, as the beams are set close enough to defeat the Hollywood fantasy of stepping over and around them. But if Hollywood stuck to the facts, those movies would be pretty short. These systems are very reliable and foolproof. The same can be said for sound activation systems. Changes in sound, such as a window breaking, or other noises associated with illegal activity, can be used to start the surveillance.

Often, the electronics for these devices are based on simple operational amplifier (Op-Amp) circuits that anyone even slightly familiar with electronic construction can easily build. The cost will normally come in under five dollars, but you will have to connect the detectors (light or sound) to your surveillance system. Even so, overall expense here is minimal, considering the peace of mind factor.

Alarm Activation

Alarm activation is usually just a matter of connecting your surveillance system to your alarm system. When the alarm is tripped, surveillance will begin. The hardware associated with this connection can be fairly expensive; however, the benefit of this arrangement may well override the cost. Some alarm units already have provisions for this type of procedure. For that information, you will need to check your alarm system's manual or dealer.

A variation on this theme couples access control with CCTV. Most higher-end equipment can be set up to automatically pan and zoom a camera to a pre-programmed position when a trigger is activated. For example, if someone swipes a card at a door, the door unlocks and the camera automatically calls up a monitor and zooms in onto the door to get a clear view of the person walking through it. You can get quite creative with this if you put a little thought into the system as it is being designed and installed.

Most of the video retailers (their names and contact information can be found in the source list in Appendix B) carry devices that couple surveillance and alarm gear. Additionally, any of your local security companies should be able to help you get set up. If they don't have what you need, they will be able to get it.

Timed Activation

The last subject regarding automatic or external stimulus of your system is activation by timer. Here, surveillance is started when a timer switch closes. There is a wide variety of timers available, and you can find them easily in retail outlets such as discount and hardware stores, and, of course, at Radio Shack. Virtually any timer will work for this purpose. What you buy often depends on the amount of convenience you want.

The more modestly designed timers simply have a dial that you use to set the time and adjacent levers that tell the unit when to activate and deactivate. With the most sophisticated units, you can set multiple start and stop times with a digital clock. There are plenty of variations in between. Again, how much you want to spend and how easy you want the whole process to be dictate which timer you buy.

RECORDING DEVICES

This is an area that is experiencing some major developments, and there doesn't seem to be any end in sight. The most obvious recording device for video signals is the videocassette recorder, or VCR, but there is considerable research going on into using compact disc (CD) and digital video disc (DVD). Additionally, investigation into techniques that more efficiently compress video information may open new avenues for using personal computers.

Digital Recording

With the leaps and bounds that are being made in the computer industry, digital video surveillance is becoming more and more advanced and refined. Currently several companies manufacture a digital video recorder that acts as a direct replacement for a VCR and can usually record thirty days or more of video. The higher-end equipment will actually mimic a multiplexer and record up to sixteen cameras simultaneously.

What they do not do well, though, is maintain a hard-copy archive of the recorded video. Images from these digital recorders can be saved on your computer, but it would take between eighty and 160 gigabytes to do so. It would take a whole lot of CDs or floppy disks to archive that video, and very quickly become very time consuming and cost prohibitive. If you only need to save specific events, rather than archive all the images captured, these devices come with software that enables you to copy specific events onto your computer and save the files in one of a variety of formats, such as AVI, DVD, Mpeg, VCD, and then view it, save it to a CD, e-mail it...anything you can do with any other computer file.

On another note, the images captured are already digital video, and that makes it a whole lot easier to do any enhancements if you need to. It also makes it very easy to capture still images and save them as standard .jpg files. Another nice feature is that these systems usually have built-in video motion detection so that they do not record areas where nothing is happening. They take video motion detection one step further though, and can save one or two minutes before and after the motion event. Unlike standard "dry" contact devices, where tripping a trigger switches on a device that captures a great shot of the back of someone's head or, even worse, a door closing, these digital systems capture the whole event. They accomplish this by continually caching the video into RAM (random access memory) until motion occurs, then moving relevant minutes of the cached memory into permanent storage when motion triggers an event capture.

You can use a personal computer as well, with off-the-shelf video cards and a little creativity. For example, ATI Technologies' All in Wonder Radeon 7200 can perform many of the tasks just described with the included software package. It takes a little practice to use, but for limited and low-cost results, it can be quite effective. Once you start playing with video capture you do have to understand the concepts of motion estimation, video codecs, bitrates, frames per second, and frame sizes, and you need a lot of patience. But if you are of a digital inclination and enjoy learning by experimenting, you can get some good results.

At present, the methods computers use to compress video are pretty inefficient, and as mentioned earlier, storing video requires almost unacceptable storage space. This has slowed the adoption of home computers for this purpose. Even CDs are limited when confronted with the vast amount of space needed to record and preserve live video signals. However, some new products that might change all that are on the horizon. A new video compression technology (codec) called mpeg-4 can take a seven-gigabyte video clip and compress it to the 650-700-megabyte range (that's small enough to fit onto a CD) and maintain DVD image quality. There is another format called Divx (not to be confused with the failed DVD predecessor DIVX) which does cause you to lose some video quality, but it does compress the video and audio quite nicely. In the near future, these formats will certainly become useful, and affordable, surveillance options.

Recording with a VCR

Now, let's look at what is available in the VCR market. For most applications, this is probably the best way to go. I think most of us are familiar with video-cassette recording and the evolution it has experienced in the last two decades. Today, VCRs are smaller, more efficient, and cheaper than ever before, and they produce far better reproduction than their predecessors.

If space is not a major consideration, then a visit to your local discount department store is probably the answer to your recording needs. Compact four-head VCRs are commonly seen in the $70 to $80 range and do an admirable job of preserving the images your camera sees. Most of the time your VCR will only be able to record the images captured by one camera. A sequencer on the VCR input enables the VCR to handle more than one camera, but as the sequencer changes from one camera to the next, you are still only recording one camera at a time. Multiplexers and quad processors can be used to split the screen into multiple camera inputs, all of which are taped by the VCR. See Chapter 8 for the details on quad processors.

The four-head arrangement on your VCR can be important if you want the ability to freeze-frame the videotape. What we are talking about here is a video recording head assembly that has four individual recording heads instead of the more commonly seen two-head construction. I say more commonly seen, because the four-head recorders are becoming more and more prevalent. You may well be able to purchase them for about the same amount as the two-head units.

Why is this important when you want to stop the picture on your VCR? The two additional heads keep the frozen image (frame) clear, as opposed to the all-too-familiar frames with static bars or groups of snowy lines. If it is necessary to get a perceptible single frame, say for identification purposes, then four heads are essential.

Now, off-the-shelf consumer recorders do not offer much in the way of versatility when it comes to the amount of video information you will be able to record. Most of them are limited to two or three speeds and a maximum recording time of about eight hours, utilizing T-160 VHS tapes. For many operations, that will be more than enough. Also, it is a simple matter to change the tape when it approaches the end of its capacity.

If you want the ability to edit your tapes, the cheap VCRs are great, as you will need two of them for that job.

Bear in mind I am talking Video Home System, or VHS, here, not BETA. BETA, due to the nature of its format, does produce a better image. Technology has vastly improved the quality of VHS over the original grainy pictures it produced when it was first released, but the BETA arrangement more closely follows the National Television Standards Committee (NTSC) parameters, and so renders a better product.

The problem is that you are not going to see BETA equipment in your trusty Kmart. Today, BETA is almost exclusively used by the broadcasting industry. The machines are available, but the cost is high. That is due to the more limited market, as well as the quality and durability that is engineered into the gear. Bottom line here, BETA is now made for TV stations and serious business use, such as advertising.

So you will likely be working with standard VHS or one of the high-end formats, such as H-VHS or VHS-C. For most purposes, these will more than do the job. In fact, when the image from a camera is documented at the highest record speed, it can be hard to tell the difference between a VHS signal and a broadcast signal.

Time-Lapse Recording

However, the story of VHS doesn't stop there. If six hours isn't enough time for your application, then there are the time-lapse recorders to consider. These come in a variety of sizes, shapes, and flavors, and many of them furnish considerable flexibility concerning total record time.

As of this writing, the time-lapse units provide between ninety-six and 960 hours of record time on a single T-160 VHS tape. 960 hours is forty days. Naturally, as the name implies, the recording is not continuous. The images are captured as individual frames at specific intervals, with time settings ranging from two hours to the full 960. It goes without saying that the longer the record time, the greater the interval between frames. Hence, the tape, when in motion, will product a very jumpy, jerky image.

For a scenario that doesn't require full motion documentation, though, forty days is nice. That definitely reduces the time you'll spend changing tapes, which you can put to good use doing more important things, like running your business.

If your circumstance does need constant attention, those clever manufacturers have that covered, too. The people that bring you time-lapse recorders also have a line of real-time units that will record up to forty hours on a T-160 tape. As might be expected, the secret of these units is slowing down the recording and the playback speed. In fact, at the forty-hour capacity, the tape is barely crawling along. This has a visible effect on the quality of the images, because the slower the tape moves past the recording heads, the worse the quality of that recording. The recordings are intelligible enough to at least see what is going on, though.

These real-time recorders are used extensively by businesses and have served them well. The point I am trying to make here is just that there is a genuine correlation between the quality of the images and the speed at which the tape is recorded. This is a fact of life that cannot be avoided, regardless of the technology you employ. The future may hold something different, but for now we just have to live with it.

Wearable Devices

All right, let's say you want to play James Bond and actually wear a video camera. And, I don't mean as a tiepin. Maybe you want to document a meeting or archive an event. You have two options. The first involves wearing a wireless device that transmits the signal from the camera, via radio waves, to a nearby receiver. Not a bad way to go, if you can stay within range of the receiver, but more will be said about this approach later. Your second option is to wear a miniature VCR connected directly to your camera. Several manufacturers make videocassette recorders that are easily concealed on your person, as long as you are wearing more than a bathing suit. These gems fit in pockets, purses, or shoulder bags; they can even be rigged to fit in the small of your back.

On the average, these micro recorders are in the six inches by four inches by one and a half inches range, which isn't much larger than many portable radios (see Figure 7.1). These units are powered by batteries, and a short cord is needed to connect the camera and the recorder. The cord is normally easily concealed, and one distinct advantage to wearing the recorder as well as the camera is privacy. Being a closed circuit system, no one else will be able to see what your camera sees—a possible problem with wireless systems.

PHOTO COURTESY OF SUPERCIRCUITS, INC. USED BY PERMISSION.

FIGURE 7.1   Miniature videocassette recorder.

Most of these micro recorders use eight-millimeter tape and digital recording, and, compared to the cameras, are expensive, usually in the $750 to $1,000 range. If the occasion calls for it, though, this is one very effective way to document events that happen around you.

CHOOSING THE RIGHT RECORDING MEDIUM

Rule number one in choosing a recording medium: Never use cheap tape! The damage it might cause to equipment and the degraded picture quality are just not worth the savings. Videotape today is so reasonable in price that I strongly suggest you stick with major brands. Forget the cut-rate stuff.

The life of a videotape depends entirely on how often it is played. A tape that you use over and over will only last a few years, while a tape that you record and then place on a shelf will store your data pretty much forever. The official line on magnetic tape storage is "at least ten years," but I have tapes made twenty years ago that are still perfectly serviceable.

The equipment to transfer videotaped images to CD or DVD is getting less expensive and more user friendly every day. Data stored on CD or DVD has a potentially unlimited shelf life, but of course, you can't record over a DVD. At least, not yet. You can buy CDs, called CD-RW, that can be used multiple times; but as I've mentioned many times, video takes up a lot of computer memory. CDs only hold 650 megabytes, and that just might not be enough room.

MONITORING WITH A COMPUTER

As you may recall, I lamented the reality that present video compression methods do not lend themselves well to using computers for preserving video images. This is due to the intensive memory requirement on the PC. There are boards that can be purchased for video capture, but be prepared to install an additional twenty to thirty-gigabyte hard drive just to handle image storage.

Furthermore, even at that, we are talking time lapse. Trying to record full motion with a computer would entail immense memory space, for just minutes of image, and at this point, it really isn't a practical approach to the task. Hopefully, the future will bring more efficient storage devices that will accommodate this type of video recording.

CONCLUSION

In this section, we have explored methods for recording and monitoring your surveillance cameras. Additionally, we talked about some of the options for remotely tripping your system. Any or all of this information may well be of value to you with either your initial surveillance setup or expanding your system as your needs change.

Again, all of the described gear and methods are easy to implement. The equipment comes with manuals that guide you step-by-step through the process of installing or adding it to your surveillance scheme.

Advanced Video Equipment and Accessories

This chapter discusses some variations to the standard video cameras and surveillance configurations. That is, I'm going to discuss the bells and whistles found on high-end cameras and observation systems that utilize more than one camera. First, though, some background on what makes today's cameras so powerful and versatile.

OPERATION OF THE CHARGE-COUPLED DEVICE (CCD) IMAGE SENSOR

Dating back to 1970, the charge-coupled device, or CCD, has revolutionized television cameras and, to some extent, television itself. In terms of picture quality, the early CCDs and metal oxide semiconductor imaging devices, left a lot to be desired. Those limitations were short-lived, though, and in 1979 Bosch introduced the first broadcast-quality studio camera using CCD technology.

In many ways, the CCD (see Figure 8.1) operates similarly to the older image tubes such as the orthicon and vidicon. In both cases, a serial-style string of data, representative of the image falling on the actual pixel matrix, is delivered as an output. This information is then sent to an amplifier and synchronization circuitry.

FIGURE 8.1 A couple of typical CCD image sensors.

The major differences between the solid-state imagers and the vacuum-tube devices lie in reliability, light sensitivity, power requirements, and durability. With the vidicon, strong light sources, such as the sun, caused the undesirable effects of smearing and blooming. Additionally, the image matrix was easily burned by such illumination, which often resulted in a damaged, or even destroyed, tube.

CCDs are not sensitive to any of these problems and, within reason, cannot be damaged by the light source regardless of its strength. Also, smearing and blooming are a thing of the past, as the solid-state approach simply is not susceptible to that problem. Additionally, the CCD image chips are far less expensive than the older vidicon tubes.

As far as durability goes, the CCD has a big jump over the tubes. Vidicons are made with the electron gun and image matrix enclosed in a glass envelope. Naturally, the gas content is evacuated as much as possible, hence the name vacuum tube. The CCD, on the other hand, is fashioned as a silicon array and is a solid structure fastened to a ceramic housing and connected by tiny gold wires, hence the name solid state. The only glass involved is a small window that allows the image to shine on the light-sensitive matrix.

It doesn't take a lot of imagination to see how much more vulnerable the vidicon is to breakage than the CCD. With the vidicon, even light jolts could result in hairline cracks that would ruin the tube. Any really rough treatment was almost certain to destroy the delicate vidicon.

The CCD, however, can take a lot of punishment before it is damaged. In fact, short of applying multiple blows with a large hammer, it is downright difficult to hurt the image chip. This is just the nature of solid-state electronics. That doesn't mean you should abuse a CCD, just that normal bumps and jars are not as likely to do harm as with the vacuum-tube imagers.

Concerning solid-state imagers, another major difference between the old and new technologies is power requirements. Vacuum-tube cameras had to develop voltages in the 200 to 400 volt range, depending on the type of tube used, where CCD and CMOS devices operate with a mere five to twelve volts. Not only is this a safer situation, it goes a long way toward simplifying camera circuitry.

As with most vacuum tubes, the image units relied on a high plate voltage for proper operation. In addition to amplification and sync generation, tube camera circuits had to supply this high voltage. Furthermore, the tube imaging devices incorporate a filament as an electron source, and anytime you have a filament, you have heavy current demand. The vidicons and orthicon were no exception to that rule. The old tube cameras were notorious for drawing inconvenient quantities of both voltage and current.

By comparison, the CCDs draw miniscule amounts of voltage and current. Twelve volts at 130 milliamps is about the norm for even color CCD cameras. It is often very practical to run many of these cameras off batteries. The days of having to always plan for a significant power source are history.

Perhaps the most significant improvement, though, is longevity. A vidicon tube has an expected lifetime of maybe 5000 hours, while a CCD's longevity is in the hundreds of thousands of hours. It is quite possible you could use a CCD camera daily for years on end and not burn out the chip. The vidicon camera, on the other hand, would have to have the image tube replaced frequently under that scenario.

With that said, let's take a look at how the charge-coupled device actually works. The image matrix is made up of hundreds of thousands of tiny light-sensitive devices called pixels. These are akin to silicon photocells and are electrically connected in a precise grid arrangement. When light strikes a pixel, it builds up a charge proportional to the strength of the light. This is a very small charge but it does serve the purpose.

Vertical and horizontal registers built into the CCD chip are used to transfer these charges off the matrix and to the other camera electronics. This circuitry includes amplifiers and digital signal processing. The transfers occur at the usual sixty-times-per-second video field rate.

Now, unlike standard cameras with mechanical shutters, these devices employ an electronic shutter. This provides a tremendous versatility in terms of exact exposure. The way this works involves the intensity of the image falling on the CCD. If the light is really weak, the chip will allow the sensor to absorb light for the full 1/60th of a second. If the light is very strong, the processing circuit cuts back on the absorption time in relation to the illumination strength.

That process results in the very precise exposure encountered with CCD cameras. Regardless of the shutter speed, the camera will still deliver sixty fields per second, but the exposure control provides an almost perfect television picture every time.

Let's make a quick comparison between CCD and CMOS image devices. Both are, of course, solid state in construction and nature. Both require a nominal amount of power, and both are very small, light, and durable. In fact, visually they are difficult to tell apart.

However, there is some distinction between the two imagers, probably the most significant of which is the construction technique employed with each device. CCDs are a single component in an overall system, whereas CMOS chips usually contain just about the entire circuit on a single piece of silicon. In a CMOS camera, you will find a few discrete capacitors or resistors, but the imaging, sync, and amplification are all done within an integrated circuit that looks remarkably like a CCD image chip.

That might make the CMOS devices seem to be the best of both worlds, and that may yet prove to be the case. CMOS is getting better all the time. To date, though, the CMOS imagers just haven't been able to produce the same picture quality as CCDs. If very high-quality resolution is a must, a CCD is what you need.

On a final note, and this has been mentioned before, CMOS chips generally use less power than their CCD counterparts. This is, after all, the nature of complimentary metal oxide semiconductor technology. As with the logic chips of early computer development, CMOS devices were always chosen over transistor-to-transistor logic (TTL) when low power consumption was a factor. And, that rule applies to this day.

That provides you a little knowledge regarding the differences between the CCD and CMOS video cameras and image sensors. Bear in mind it is only a small part of the total picture, but to set up your own surveillance system, you don't really need to know even this much. This falls more into the category of interesting trivia, unless, of course, you are an electronic engineer developing a new solid-state TV camera design.

NTSC VERSUS DIGITAL FORMAT

Much of this information has at least been touched on before, but is presented here again to make this discussion coherent. There really isn't a lot of difference between NTSC and digital cameras as such. Both employ CCD or CMOS detectors that will produce a serial-type output. What makes the difference between digital and analog is what happens to the image after it leaves the semiconductor registers. With digital cameras, the image is presented as a serial string of information that can be employed by a computer with the right software installed. With analog cameras, the camera must also apply timing and synchronization to the signal, and through a process called *digital signal processing* (DSP)—a bit too complex to cover in this text—must define the color elements of the picture. Additionally, the analog camera has to arrange all this information into a standard NTSC format so the monitor will understand it.

The timing of all this conversion and transmission of data is crucial to proper operation. The TV/monitor does the rest of the work once the signal leaves the camera, but that "work" is, in effect, decoding the various signal components developed by the camera. Video is complex and very temperamental and it doesn't take much to upset the timing. (Take a look at the timing diagram of a single line of NTSC video, shown in Chapter 5, and you'll notice that we're talking timing tolerances of microseconds.) The timing can be thrown off by a variety of factors, including power line glitches, bad or failing electronic components, external influences such as strong magnetic fields, and the list goes on and on.

Analog Cameras (NTSC)

The National Television Standards Committee (NTSC) established the standards by which analog television (standard commercial broadcasting and the signal most observation cameras generate), has operated since its inception in early 1953 in the United States. Other formats, such as PAL and SECAM, are used in Europe and elsewhere in the world.

Some people say that NTSC really stands for "Never the Same Color" or "Never the Same Color Twice." Admittedly, there are some problems with the system. Synchronization of the various components of an NTSC composite signal is crucial, allowing for almost no forgiveness in terms of timing. The person who observed that the three most important aspects of video are timing, timing, and most of all, timing was right.

Some of the criticism directed at the NTSC is valid. But in all fairness, coming up with those standards was a monumental task back in 1951, and in only two years the NTSC came up with a format that has served us for almost five decades. That is an impressive track record, even with the problems. The new digital and High-Definition Television (HDTV) formats put on a great show, of course. This is admirable technology that will unquestionably replace NTSC, probably in the not-too-distant future.

Anyway, back to the issue at hand. The NTSC format produces a composite signal. This means that all the information your television needs to render a proper picture is being sent in a single line. Here is where some of the timing is involved, as the individual components of that composite signal have to be interlaced to be useful to the TV. Obviously, you can't send all the data at the same time. For more information about exactly how this signal is broken down, see Chapter 5.

Precise timing is needed to manage the task of getting the right signal to the TV at the right time. The content of the signals is far less important than the fact that they arrive where they should on time. These signals include the video information itself (luminance), the color information (chrominance/color burst), and the synchronization (sync) signals that control both vertical and horizontal image positioning.

When everything is working, you get a great picture on your television screen. When it is not, well, the picture won't be so great. But that is true of all TV formats. PAL (most of Europe) and SECAM (France) are just as sensitive to proper timing as NTSC, and so are the newer HDTV and digital modes.

NTSC uses an interlace system to display the information on the TV set. This means that only half of the screen is actually scanned. Each of these half-scans is known as a field and there are, of course, two fields to each frame displayed. The reason behind this approach goes back to the technology of the early 1950s. The phosphorus material used in the cathode-ray tubes (CRTs) at that time had a very short persistence, or length of time it would glow. If the entire screen had been scanned each time, the top would have begun to fade before the scan reached the bottom area of the picture tube.

This problem was addressed by the committee and interlacing was adopted as the best solution. So, the first scan covers all the odd-numbered lines (1, 3, 5, 7, 9, etc.) and the second scan hits the even lines (2, 4, 6, 8, etc.). Also, to help with image positioning, the first scan starts at the upper left-hand corner of the screen and proceeds to the middle of the last odd-numbered line. The second scan starts at the middle of line two and ends up at the bottom right-hand corner of the screen.

This might at first glance seem a little weird, but the entire scan is composed of 525 lines. Since you are only scanning half of them in each field, only $262^1/_2$ lines are scanned. Thus, it is necessary to break up that odd line into one half of a line in each field. That pays off, though, in another way. Because the starting and stopping point of each field is different, the TV knows which field is being scanned and will apply the appropriate video information in the appropriate place. That way what you see will actually look like what the camera photographed instead of a jumbled mass of squiggly lines and noise.

Additionally, the NTSC system uses a horizontal sync pulse of 15,743 hertz and a vertical sync pulse of 59.94 hertz. You will notice the vertical pulse is very close to sixty hertz. Actually, in the original "pre-NTSC" arrangement, the vertical pulse was sixty hertz, which was done in the interest of more pragmatic timing. Instead of having to incorporate another sync generator in the TV circuitry, the frequency of the 120 VAC power line is used for the vertical sync pulse.

This was a clever idea that worked well with black-and-white, but gave the NTSC some trouble when it came to color television. So the NTSC was forced to change the vertical pulse just slightly to the 59.94-hertz rate. But that change was enough to require an additional sync generator circuit for proper timing.

The list of NTSC parameters runs almost a page and a half, and covers everything from the actual time in microseconds each field and frame should take to such things as the lower video sideband and horizontal blanking. But for our purposes, what we have covered is all you really need to know to understand the NTSC format and the difference between it and digital TV. There is some interesting information in the complete format specifications, if you are interested in that sort of stuff, and the specifications can be found in most sources that deal with television and video. At the risk of sounding presumptuous, you might try my book *The Video Hacker's Handbook* (Delmar Publishers, 1997, ISBN 0-7906-1126-0).

Digital Cameras

Let's move on to digital cameras and the approach they use to display information on your PC or dedicated digital monitor.

There isn't a whole lot to be said about digital systems. Being digital, all data is transferred to the computer or monitor in a serial string. That is, the individual pulses from the digital image sensor are sent out one bit at a time. They are usually sent through a video amplifier first, to increase their strength, but once that happens, the computer does the rest of the work, as mentioned earlier.

That is why video cameras advertised for use only with computers or dedicated digital systems are somewhat cheaper than NTSC units. All that is required is a CCD or CMOS imager, a video amp, a power source, and a cable to transfer the serial string to the computer or monitor.

As you may recall from Chapter 2, charge-coupled devices (CCDs) and Complimentary Metal Oxide Semiconductor (CMOS) image chips produce just such a digital output. However, with NTSC cameras, additional electronics are necessary to provide the required analog signals for proper operation. Of course, additional electronics means additional cost and a higher price tag.

Having cameras produce digital output that is translated to an analog signal is actually going to be a benefit to commercial broadcast stations. As of this writing, all television stations must be able to broadcast their programming in both analog and digital (simulcasting) by the year 2004. I expect that deadline will be pushed back, but for the time being, that is the plan. This, of course, is going to place a huge expense on the individual television facilities, as everything has to be converted to digital format. That is in addition to maintaining the already existing analog gear. By the year 2012 or so, all analog broadcasting will vanish and everything will be strictly digital. This move is supposed to vastly improve the quality of the television pictures sent to you.

There are some implications here for long-term planning for your security system. Will the TV or monitor that you buy tomorrow still be useful after 2012 (in the unlikely event that it's still running in ten years)? Should you be thinking along the lines of succession planning, and keeping in mind when it would be a good time to switch to more digital equipment? The implication is that analog systems will be gone in the future. Whatever analog equipment you do have will still work; you just won't be able to replace anything that stops working. Planning ahead is tough at this point as very little digital gear is available and there is no definitive schedule as to when the digital stuff will be available.

How will the switch to digital affect the crowding in signal space? Will it get worse, or will there be more room for "amateur" broadcasting? Digital will have little effect, if any, on band usage, and I seriously doubt that the FCC is going to allocate additional space for amateur TV.

Now comes the good news for the TV stations, however. Since many of them are presently using the more advanced CCD studio cameras, and these cameras are already producing a digital signal, all that has to be done is to tap off at the CCD output, and they will have the beginning of their digital system. Naturally there will be a lot more to getting on the air digitally than that, but it is a good start.

For the present, our analog NTSC cameras will do the job we need them to do.

ADVANCED CAMERA SYSTEMS

The fancier video cameras are a good place to start, if you want an upgradeable system. It is hard to add much to even the most basic of the board cameras. They already have automatic shutter adjustment for a wide range of lighting conditions, and automatic gain control. And with the color devices, automatic white balance is standard. So, where do you go to add features to these marvels?

One of the most popular additions is automatic back light control (BLC). With this innovation, the camera will adjust for a situation where the light is considerably stronger behind the primary subject. Without BLC, when a subject is strongly back-lit it appears dark, due to the camera accommodating the entire scene. You lose detail in the important part of the picture, the primary subject. With the back light control, the background washes out as the camera compensates for less light on the subject and sets the shutter accordingly. There are times when you will really love this feature!

Another available feature is the C-Mount lens. While most of the board cameras come with a built-in lens, the larger box cameras often require a lens to be mounted to the camera. I say most board units, as some models do offer an auxiliary lens version. In almost every case, the lens to be added will have "C" mount threads that mate with a special mounting ring on the camera. Thus, these lenses are referred to as *C-Mount* lenses.

Now, you might be asking why you would want this feature. Why not stick with the cameras that come with a lens and not incur the additional expense of adding a lens? The answer is that you may not need this feature. Often, the lens that comes with a board camera will be more than adequate for your application. However, there are going to be other times when you will need versatility in optics. There may be special situations where the included wide-angle lens normally supplied with cameras will not work. It is in these circumstances that lens selection, and C-Mount lenses, are handy.

I will say more about the vast assortment of lenses available later in this chapter. For the time being, let me say that the ability to change camera lenses increases the flexibility of virtually any surveillance system. Longer lenses for tighter shots, super wide-angle lenses for extreme area coverage, and zoom lenses that allow you to change the area of view are just a few examples of the kind of flexibility I'm talking about.

Another nice add-on is automatic iris output. This property is usually found with C-Mount cameras, and it allows you to use external lenses that have either DC or video-driven iris control. This is a nice feature, as it enables you to adjust the iris in the lens manually or remotely. That, of course, permits you to adjust the amount of light that reaches the CCD or CMOS sensor.

Resolution is an additional factor that often improves with higher-end cameras (see Figure 8.2). Many of these devices have resolution in the 420-line range for the color cameras (remember, 330 to 380 is the usual resolution for color cameras), and I have seen one monochrome unit that has 570 lines of resolution (380 to 410 is standard for monochrome). That is better than a standard television set, which displays 525 lines. Naturally, the better the resolution, the better, clearer, and more detailed the picture.

This has been a quick look at why you might want to purchase one of the more advanced cameras for your system. Much depends on how you plan to set up and use the cameras. One thing is for sure; today's video devices provide an almost endless selection of equipment. That allows you to fine-tune your surveillance and get exactly what you want.

PHOTO COURTESY OF SUPERCIRCUITS, INC. USED BY PERMISSION.

FIGURE 8.2 High-resolution camera.

MULTIPLE-CAMERA SYSTEMS

Next, let's take a look at multi-camera configurations (see Figure 8.3). For many observation arrangements, a single camera does the job, but other scenarios require two or more devices to provide proper coverage. This is where a few new accessories come in handy.

As the name implies, these are surveillance systems that employ a number of different cameras, usually aimed at different areas of your home or business. The obvious advantage here is the complete coverage that you can obtain. At one time, it would have been necessary to monitor each camera separately using a separate screen for each camera. Actually, this is still done in some high-end systems, such as gambling casinos and large business operations. For most small-business and home applications, though, a single monitor with the screen image split into several sections is more than adequate.

PHOTO COURTESY OF SUPERCIRCUITS, INC. USED BY PERMISSION.

FIGURE 8.3 A complete and ready-to-install four-camera multi-camera system.

In this arena you have several ways to go. Probably the easiest and cheapest approach is the *quad processor*. Next in line would be the *multi-record processor*, and rounding out the selection is the *multiplexer*. Actually, the multi-record processor and multiplexer are very similar in terms of features and price.

In an effort to explain each of these systems, I will cover what they do and what can be done with them in a surveillance system. Let me address the quad processor first.

Quad Processors

Quad processors are devices that split the monitor screen into four separate views, which accommodates up to four cameras. The views are real time in nature, thus the refresh rate is thirty frames per second. That is the same as any

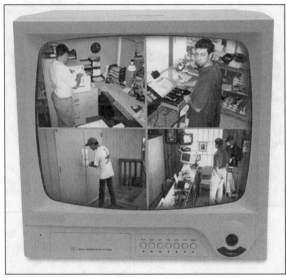

PHOTO COURTESY OF SUPERCIRCUITS, INC. USED BY PERMISSION.

FIGURE 8.4 The monitor view of a typical quad processor running 4 cameras.

analog monitor or television set used to observe camera vistas. Normally, these processors provide dual outputs and adjustable gain on all four channels (inputs) (see Figure 8.4).

Some of the more expensive units will allow for more than four cameras. Usually in multiples of four, these processors divide the camera views into pages of four scenes each. With eight cameras in the system, you would have two pages, each with four camera views. This is not as convenient as the multiplexer approach where all views are scene on a single screen, but for the economy system, it is a good compromise.

Multi-Record Processors

The multi-record processors normally allow up to sixteen camera scenes to be viewed at any given time. They also permit those scenes to be recorded on a single videocassette recorder. For archival purposes, this innovation is downright handy (see Figure 8.5).

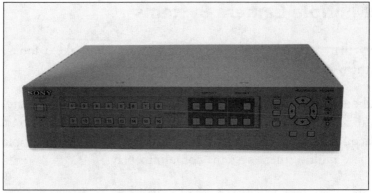

PHOTO COURTESY OF SUPERCIRCUITS, INC. USED BY PERMISSION.

FIGURE 8.5 A typical multi-record processor.

Additionally, most of the multi-record processors offer features such as auto switching of all cameras, built-in motion detection, tape tagging, built-in time, date, and title generator, video zoom, digital freeze frame, external alarm inputs for active areas, automatic switching to full screen when alarm is activated, auto color phase and synchronization, monochrome/color combination views, on-screen character display, and even other goodies. As you can see, the multi-record processor is somewhat more complex than the quad processor, and this factor does justify the added expense.

Multiplexers

As said before, the last device is similar in nature to the multi-recorder. Multi-plexers can also view and record up to sixteen images on a single monitor and VCR. They too have the various features like freeze frame, external alarms, on-screen character display, auto alarm switching to full screen, and most of the other highlights of the multi-recorder. The only additional features found on many multiplexers are computer access ports and on-screen menus for setup.

All in all, there isn't a dime's worth of difference between the last two devices. Price may well be the deciding factor when purchasing such a system.

Using Multiple Camera Systems

Okay, now that we know what each unit will do, let's take a look at how to integrate them into a surveillance operation.

For just about any circumstance where surveillance is desired, more than one camera is an asset. As always, there are exceptions to this rule, but on the whole, you will acquire better protection and greater security with multiple cameras. That is where multi-camera systems come into play.

For example, you may want to safeguard your home with a camera at the front door, one at that blind spot in the backyard, another looking out your driveway, and a fourth camera in the children's room. Installing the cameras is a simple enough matter, but viewing all four will require either four separate monitors or a quad processor.

Of course, you could use a multi-recorder or multiplexer, and might well want to invest in such a device if you plan to expand your system down the line. But, for the basic coverage I described above, the quad processor will do a terrific job. You will be able to see all cameras at once without four space-consuming, and expensive, monitors. A quad processor shouldn't set you back more than $300, even if you buy a multi-page unit.

With scenarios that require ten, twelve, or sixteen cameras, such as a large home or small- to medium-sized business, the multi-recorder or multiplexer does a much better job. Not only do they cut down on the monitor count, but they also provide full VCR capability. And don't forget about the raft of other features these devices offer.

With large systems, record capability is very desirable. It is hard to view sixteen different scenes without missing something, and if Murphy's Law applies (which it usually does), you will miss the scene you really needed to see. The VCR, on the other hand, doesn't miss anything, and it preserves everything for you on tape.

In addition, the ability to tag tapes with such things as the date and time of the recording will prove invaluable. This is especially important if you plan to retain and/or edit the video recordings. Other benefits such as the alarm features and frame freezing will enable your system to better protect the property it is designed to protect.

There you have it—a brief overview of some of the advanced cameras and viewing aides available for your surveillance setup. As with just about everything, certain factors have to be weighed against cost. However, when those factors include the safety of your family, employees, or yourself, a little additional expense may be justified.

WIRELESS CAMERA SYSTEMS

We have touched on this subject from time to time; hopefully this discussion will give you a better understanding of what this is all about. The term *wireless* in this context refers to a surveillance system that uses radio frequency (RF) energy to convey the video, and sometimes audio, information to where it is needed. Simply put, radio transmitters are used to send the signal to a receiver.

That sounds elementary enough on the surface, but there is a little more to it. For example, RF transmitters come in a variety of types, sizes, power outputs, and capabilities (see Figure 8.6). Some require an FCC license to be operated legally, while others fall under the umbrella of low-power devices that do not need to be licensed. When they do need to be licensed, this may be in the form of an FCC permit usually issued yearly at a cost to the licensee. Transmitters may also fall under the auspices of the Amateur Radio Service (ARS), where the license is earned by passing various tests. Which license you need will depend on factors such as the frequency being used, the power output of the transmitter, and the purpose for which the signal is being transmitted.

When a transmitter is covered under the Amateur Radio Service, for example, signals cannot be radiated for commercial or business purposes. It is all right to use that frequency and equipment to protect your home, but not your business, as your business qualifies as a commercial operation. The ARS, or HAM radio, provides a wide range of frequency allocations for video signals, so it is a good prospect for home surveillance systems. But you must always remember that it cannot be used for anything but amateur purposes.

> **N O T E**
>
> The terms *ARS* and *HAM radio* are synonymous. ARS stands for *Amateur Radio Service*. *HAM radio* is a slang term dating back to the early part of the twentieth century.

Commercial licenses, on the other hand, are very restricted in terms of what can and can't be transmitted. There are very few allocations for television use, because the 6 MHz bandwidth or fast scan images that television requires is at

PHOTO COURTESY OF SUPERCIRCUITS, INC. USED BY PERMISSION.

FIGURE 8.6 Two versions of the trusty low-power video transmitter. The unit in the foreground is VHF; the other is UHF.

a premium among the commercial radio services. As of this writing, most of those frequency allocations are in the microwave region of the radio spectrum.

Additionally, you have to renew a commercial license periodically, usually yearly, and pay a fee to the FCC each time you renew. With amateur radio, once a license has been earned, it only has to be reregistered every ten years to be valid. However, as the name implies, a commercial license is for commercial, or business, use. The FCC issues commercial licenses only for business purposes. The cost varies widely depending on the type of use (police/fire department, taxi company, newspaper dispatch, etc.) and the complexity of the system. Also, these fees can change virtually without notice. Amateur radio licenses, while far more frequency-versatile, are strictly for nonbusiness use. The two types do not interchange!

As for the low-power device category, this equipment is very user friendly. That is, it does not require any FCC sanction other than the designation low power. It can be used for both personal and commercial purposes, and as long as it doesn't interfere with the operation of any licensed station, it is entirely legal.

The big drawback is this bit about low power. Depending on the transmit frequency, these devices are allowed to emit a maximum of 100 milliwatts down around the AM broadcast band and only a few milliwatts up in the microwave region. That translates into a relatively short transmitter range. Often, the best these devices can do is only a few hundred feet, even in line of sight (see Figure 8.7).

For many applications, though, that is all that you need. Sometimes, short range is actually an asset, as the shorter the distance a signal carries, the fewer receivers it reaches. As long as you are within range, the transmitter is doing what you need it to do. And people you don't want receiving your signal are out of range and can't receive it. Think about it. It makes sense.

Now that we have the ground rules, let's take a closer look at this equipment. There is a veritable smorgasbord of stuff to choose from out there; what you buy just depends on what you need the equipment to do. So, let me go through the three components required for any wireless application, the transmitter, the receiver, and the antenna, and cover what is available.

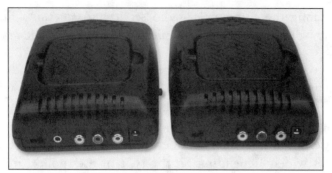

PHOTO COURTESY OF SUPERCIRCUITS, INC. USED BY PERMISSION.

FIGURE 8.7   A 2.4 gigahertz video link with separate transmitter and receiver units.

Transmitters

When it comes to transmitters approved for the Amateur Radio Service, a number of companies offer gear. Some of it is in kit form, while others are open boards that require a case. Some of these devices are so small you can hold them in the palm of your hand and that is including a color camera. There are also completely assembled and tested transmitters and transceivers available.

Due to the wide bandwidth of fast scan, or live television signals (six megahertz), the seventy-centimeter band (420 to 450 megahertz) is the first place you will encounter television in the amateur radio spectrum. There simply isn't another band below these frequencies that has the room. *Fast scan* refers to the "live" version of image transmission, as opposed to slow scan. The other version of transmitted TV (and the term *TV* is used loosely) is *slow scan*, in which only still pictures can be sent. By *live*, I don't mean what most people think of when they hear "live television"—shows that are broadcast from a live performance rather than taped. *Live* refers to the type of television we all know and love—that is, what we watch at home. The "live" part comes from the fact that the images are moving and appear to be alive.

On seventy centimeters, you will find three channels designated for amateur television (ATV) use. They are 420 to 426 MHz, 426 to 432 MHz, and 438 to 444 MHz. In my exploits with ATV, I have normally used the 438 to 444 channel, but that is more a matter of personal preference than anything else. All three will work great.

The next band for ATV is thirty-three centimeters (902 to 928 MHz). Here, two channels have been designated; 909 to 915 MHz and 921 to 927 MHz. As you may have noticed, the higher the band, or frequency, the larger the spectrum. With thirty-three centimeters there is twenty-six megahertz of space, while 2 meters has only four megs and twenty meters has only 200 kilohertz. Neither of the latter could accommodate ATV.

The twenty-three-centimeter band (1240 to 1300 MHz) is next, and it has three allocations; 1240 to 1246 MHz, 1252 to 1258 MHz, and 1276 to 1282 MHz. The area of 1288 to 1294 MHz does allow ATV simplex, but you will have to share the space with wideband experimental stations.

As you move up the spectrum, the equipment becomes harder to find or build, and also becomes more temperamental. This is due primarily to the characteristic of those frequencies. For example, what might be a short jumper at two meters is an entire coil at twenty-three centimeters. Also, in these high-frequency ranges (microwaves), the printed circuit board (PCB), with its parasitic inductance and capacitance, actually becomes a part of the circuit.

From here on up into the microwave spectrum, there is plenty of space for ATV use. Since amateur radio has operating authority all the way to and beyond 300 gigahertz, there is no end to the possibilities regarding what frequency to use. The major problem is, again, complexity of the equipment.

One last caution about using the amateur radio bands involves the necessity, as dictated by the FCC, to identify the station no less than once every ten minutes. This means putting the station's call sign out over the air within that timeframe. With ATV, it is best to transmit a video image of the call, as well as a voice message. It also doesn't hurt to send the call out in Morse code as part of the voice signal.

There are any number of ways to do this, from very simply shooting a graphic view of the call with a video camera and speaking the call sign into a microphone, to more elaborate equipment with built-in voice modules and timers. One method I have used and found acceptable is to build a pseudo video slide projector with a voice module for the phone side.

The projector is constructed from the frame of what began life as a thirty-five millimeter photographic bellows. On one end, I fashioned an assembly that will hold thirty-five mm slides and also illuminates them from behind. The other end holds a small color video camera that has been refocused for the distance between the two sections. In this way, I can transmit whatever slide has been placed in the stage section (see Figure 8.8).

Using one of the commonly available voice module chips, you can record twenty seconds of digital sound. That is more than sufficient to hold the identification message. Both the video and voice signals are fed to RCA jacks that, in turn, are used to send the signals to the ATV transmitter.

One nice aspect of this arrangement is that you can change the ID slide at will. Or, if you employ the ATV system in some other activity, specialized slides can

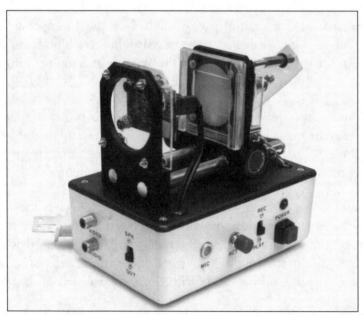

PHOTO COURTESY OF SUPERCIRCUITS, INC. USED BY PERMISSION.

FIGURE 8.8 A home-brewed unit used to provide video/audio identification of an amateur radio TV station. A small color camera captures the video from a color slide, and the audio is provided by a voice chip integrated circuit.

be made to accommodate whatever project you are involved in. Also, when size is a consideration, the unit is quite compact and portable.

With this arrangement, I merely push a button when I want the identification to go out, and both a video image and voice ID, complete with Morse code, are transmitted. I could also add an intervalometer set for slightly less than ten minutes to the system that would automatically send the signals when needed. Here again, your own personal preferences, needs, and ingenuity play an important role in how far you might go with this.

If you hold a technician or higher amateur radio license, that world is open for you to use in protecting your home. But not everyone holds one of those licenses. Not to worry! The FCC understands this fact and has taken your interests into consideration. As I said before, frequency demand on the commercial bands is so high that there is virtually no room for live television. This fact precludes using any of the four Citizen's Radio Service bands (also known as just *Citizen's Band*) for TV purposes. They are primarily for voice signals, with some data channels allowed. However, the data channels are only for narrow bandwidth systems.

Why would you need a technician or higher license—isn't the lowest level of licensing enough to get going? There are three levels of licensing for amateur radio: technician, general, and amateur extra (lowest to highest class, respectively) and the technician level is all you will need. However, the general and extra class will also allow you to use television on the amateur radio bands. The Citizen's Radio Service band is not wide enough (just about 1 megahertz wide) to accommodate the six megahertz of space that all fast scan television signals require. (*Signal* is just another term for transmission.)

Thus, except for some extremely rare scenarios, ones that might involve preserving life or national security, television on the commercial bands is taboo. Do not despair. There are some alternatives for home and business surveillance.

That brings us to the third category, low-power devices. Here, the FCC has borrowed frequencies from the Amateur Radio Service and put them to use in a shared capacity. That is to say, both the ARS and the low-power equipment use the frequencies in question. One caveat to all this is that the low-power gear cannot interfere with normal ARS-licensed operation. But the expanse of spectrum in the bands capable of handling live television, especially in the microwave range, is so extensive that usually this will not be a problem.

The equipment I'm talking about functions in the seventy-centimeter, 900 megahertz, thirty-three-centimeter, and twenty-three-centimeter bands. It comes in a variety of sizes, shapes, and configurations, but all these devices have one thing in common: low power output, usually in the single-digit milliwatt range.

Signal carry is therefore relatively short (100 to 300 feet). However, there is some newer gear that will carry about 800 feet, and if that still is not good enough, special antennas can greatly increase the range, sometimes as far as 2400 feet. The addition of a highly sensitive receiver to the system also does wonders, but we will talk more about that in a minute.

Back to the transmitters. There is some great stuff here. One company offers a combination color CMOS camera, 900 MHz transmitter, and antenna all in a package that is less than one cubic inch. Range on this one, which also comes in ATV, will vary, but should be, for most applications, in the 300 to 500-foot range. Additionally, the unit weighs less than an ounce and will operate for about eight hours on a nine-volt alkaline battery.

Another small unit measures in at 0.99 inches by 0.68 inches by 0.21 inches, transmits on either 434 MHz or 900 MHz (two different models), and weighs less than a quarter of an ounce. Operating off a nine-volt battery, range is 1500 feet. This one is incredible, but a bit pricey.

A third, higher-powered unit operates in the 2.4-gigahertz frequency range (twenty-three centimeters) and claims signal carry up to two miles. A little larger (3 inches by 1.95 inches by 0.65 inches), this transmitter puts out around 100 milliwatts and can be configured with special antennas for even greater range. If you need extended reach, this will do it for you.

Many dealers offer combination units similar to, but larger than, the camera/trans-mitter combo discussed above. I talked about this earlier, but let me expand that dialogue to this subject.

Overt surveillance cameras with built-in 900 MHz, 1.2 GHz, or 2.4 GHz transmitters are becoming ever more common and inexpensive. These allow for far easier system installation, as connecting cables are kept to a minimum. Range is usually about 300 feet, which will handily cover most homes and small businesses.

Additionally, video links are available to transmit the signals. These often include *down converters* that take the high frequency signal and transform it to standard television low VHF channels two, three, or four. In that fashion, your regular TVs can be utilized in your surveillance system. Again, these links function in the 900 MHz, 1.2 GHz, and 2.4 GHz bands, which helps make the signals clear and void of interference.

The video links are normally seen as two-unit arrangements of the rounded-box variety. One box is the transmitter, while the other is the receiver, or receiver/down converter. They usually come with a plug-in wall power module, but some can be operated on battery power. If portability is important, for instance to transmit a signal from your camcorder, then battery operation is a feature to consider.

Of course, with the aid of a suitable plug, one that matches the power input of your link, you could also power the unit from a separate battery pack. That should furnish longer transmit times and better reliability. All of this just goes to illustrate the versatility of much of the new video equipment. Considering the size, weight, and performance of today's gear, there isn't much you can't do with it.

Camera/monitor combinations are another popular item these days. I have one unit that is remarkably small (five-inch screen), light weight, and works like a trouper. This one operates in the 900 MHz band, and the monitor has a built-in link receiver. The transmitter is small and allows for infrared illumination, and both units can be battery operated (see Figure 8.9).

It has been a real lifesaver when it comes to a quick and simple on-the-spot or emergency surveillance system. Range is 100 to 200 feet, which has proven to be more than adequate for the applications I have needed it for. Similar dedicated combos come in the 1.2 and 2.4 gigahertz bands. If you haven't had an opportunity to look at one of these systems, I highly recommend you do so. I think you will like what you see.

So much for the transmitters, even as fascinating as they are. Let's take a look at some of the specialized receiving equipment and what it can do. As I mentioned,

PHOTO COURTESY OF SUPERCIRCUITS, INC. USED BY PERMISSION.

FIGURE 8.9 A typical 900 MHz B&W wireless surveillance system.
Notice that the receiver incorporates a dedicated monitor,
and the camera sports a ring of IR LEDs.

if extended range is necessary, often high-sensitivity receivers are the avenue to explore. The addition of one of these gems will do wonders for your surveillance capabilities.

Receivers

In much wireless equipment, especially the stuff we have been talking about, the receiver is an inherent part of the system. Inherent, but dedicated. That is to say, the characteristics of the receiving section of the system cannot be significantly changed. What you see is what you get. In other words, if you get a combo transmitter/receiver deal, then for the most part you can't really upgrade to a more sensitive receiver. There may be exceptions with some very expensive systems but I'm not aware of any such systems.

For many applications, this is more than sufficient. The range needed for the transmitted signal, or the capability of the transmitter itself, is well within the receiver's ability to detect radio frequency signals. However, in scenarios where extended range is desired or necessary, often the best solution is to upgrade your receiver and/or antenna. This way, the extra signal carry can be accomplished without having to increase the power output of the transmitter(s).

With that said, here is a discussion of some of the receiving gear available at reasonable cost. I realize that reasonable is a relative term, but most of this stuff has dropped dramatically in price from what it cost just a few years ago.

Like the transmitters, most of the receivers are tuned to the seventy centimeter, 900 megahertz, 1.2 gigahertz, and 2.4 gigahertz bands. With FCC allocations already available in these spectrums, they are natural choices. The key to receivers is sensitivity, or their ability to detect very small signals in the air around them.

Basically, receivers are rated in terms of sensitivity by the smallest discernible signal that they can detect in terms of negative decibel ratings per microvolt of power at certain distances between the transmitter and receiver. There will be a sensitivity rating in the specifications of better receiving gear, but not all manuals have that measurement. In short, there really isn't a number for most consumers to look for, as the sensitivity also relies on the number of interfrequency stages that the receiver employs and also the type of detection circuitry used.

Receivers come in a variety of sizes and shapes, with most being configured as small flat boxes. A suitable whip antenna can be attached, or a more elaborate external antenna system can be connected via coaxial cable. Naturally, the better the antenna, the better the receiver will work. We will talk about antennas shortly.

I discussed the portable link systems in the section on transmitters, but they deserve to be rementioned, as one half of a link unit is made up of a high-sensitivity receiver. In many of these link systems, a downconverter is used to convert the received signal to a frequency that a standard television set can handle. The downconverter merely converts say 2.4 gigahertz to a frequency that your TV can tune.

Usually, low very high frequency (LVHF) TV channels two, three, or four are selected. This, of course, makes setting up your surveillance system far easier, and cheaper. In this fashion, you will be able to employ existing television equipment and not have to purchase expensive dedicated video monitors. Also, with the flip of a switch, you can go back to normal commercial programming.

Depending on the complexity of the receiver, some of the other variations also employ downconverters. There is some high-end gear that utilizes odd-ball frequencies or downconverter channels, but this is not the type of equipment you are likely to encounter for a home or business surveillance system. For one thing, it is very expensive, and for another, a lot of this gear is restricted to use by police or government agencies.

Another commonly seen configuration is the receiver/high-performance antenna combination. Often these look like a thick cover for a hardbound notebook—about 15 inches square by 1 inch thick, which makes them, well...flat.

As for increased transmit carry, that depends largely on the frequency being received. For example, you can expect ranges of up to 1500 feet with the seventy-centimeter and 900 megahertz equipment, while you can achieve 2500 feet to as much as a mile in the 1.2 and 2.4 gigahertz bands. These distances are fairly consistent for all of this receiving equipment. As with almost everything, the more you pay, the better the gear.

The cool part is that there are no limits on the kind of receiver you can use, other than what I mentioned earlier about high-end odd frequency stuff. It doesn't matter whether you're working under an FCC license, ARS, or low power—no license is needed for receivers.

There is a quick look at receivers. You may well find even more exotic stuff in your search for surveillance gear. This is a field that is changing at an alarming rate, so don't be surprised at what comes your way.

Antennas

Talk to just about any HAM (amateur radio operator) and you will hear phrases such as "black magic" and "a science all their own" applied to antennas. This is because a tremendous amount of time, effort, thought, and money has gone into the research and development of antennas and antenna systems, and still a lot is not known or understood about how they work.

One thing is for sure, though; the people involved in this field have learned how to vastly improve signal reception, even if they don't, in some cases, quite understand the principles behind the antennas. Now, I don't mean this to sound like they are a bunch of experimenters without direction. That is not the case, as most know as much about the topic as can be known. However, this is a mysterious area that can be both frustrating and rewarding.

Now that I have set the stage, let us talk about what is available for your surveillance needs. Hopefully my introduction will help you better understand why there is such a wide variety of antennas out there. Personally, I can't think of any area of electronics where more diversity exists.

Radio frequency signal radiators for the transmitters we have discussed are all going to be fairly small. This is due to the frequencies they are designed to operate on. When I refer to the seventy-centimeter band, I'm referring to the length a full wave antenna will be. For example, a seventy-centimeter antenna will be seventy times 1/100 of a meter, or about 2.1 feet. Down at the AM broadcast band, you are talking 170 meters, or approximately 510 feet (170 x 3). You can see that as the frequency increases, the length of the antenna decreases.

The flexible whip antennas, also known as heliflex or rubber ducks, range from about six inches to as little as two inches, depending on which band they service. This fact makes them convenient antennas, as they are not obtrusive, and their flexible nature makes them hard to damage.

However, they are omnidirectional (radiates in all directions at once) Heliflex radiators are not the most efficient antennas, but they are relatively cheap, usually consisting of a long piece of wire that has been wound in a loose coil around a flexible shaft. The finished product is then coated with a rubber material that keeps them pliable. For short range, they are relatively popular.

You might think that antennas just accept signals from all directions. But transmitter antennas radiate; if they're omnidirectional, they radiate in all directions at once (360 degrees). If the antenna is used only for receiving, it doesn't radiate. However, many antennas are used to both receive and transmit.

Thinking about those stubby little antennas we have on our cell phones and cordless phones? Heliflex antennas are flexible, and most cell phones don't have flexible antennas (they don't really need to be flexible as they're so short). However, if your phone does have a flexible antenna, then it is a heliflex. When I say *flexible* I mean an antenna that can be bent out of shape and it will bounce back by itself.

If you want a more efficient radiation assemblage, you will need a directional antenna, or beam. For our purposes, these come in a wide variety of sizes, shapes, and flavors. In appearance, they will range from the Yagi-style beam to parabolic dishes to flat-pack designs. Yagi beams consist of a long metal tube known as the boom with a number of shorter tubes placed across the boom at a ninety-degree angle (see Figure 8.10). These are quite directional and provide a good deal of gain, or performance.

Parabolic dish antennas have become more common in recent years, as the cable TV industry makes good use of them. That is, both at the company location and at home receiving sites. As their name implies, they are usually a round bowl shape that concentrates the received signal at some point out in front of the dish. The necessary signal detection gear is then positioned at that point where the signal concentrates. These, too, have high gain and directionality.

You've probably noticed that satellite dishes vary a lot in size. The bigger the dish, the more signal it will reflect to the detector unit positioned near the middle of the dish.

One variation to this design uses wire mesh instead of a solid surface for the dish. These sometimes look like the rack off your outdoor grill, but surprisingly they do a very good job. I say surprisingly as it seems like most of the signal would just pass right through the open spaces between the metal rods. Apparently it doesn't.

Another modification of the dish antenna comes in the shape of a square box with a rounded face. These look something like an oversized version of the passive infrared (PIR) detector units. You know, those curved boxes on motion detector outdoor floodlights and the like. These antennas also do an excellent job of consolidating the signal and producing high gain.

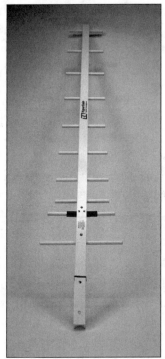

PHOTO COURTESY OF SUPERCIRCUITS, INC. USED BY PERMISSION.

FIGURE 8.10 A Yagi-style antenna.

Our last variety of antenna is the *flat-pack* beam. This is a name that I have given these gems to makes things a little less complicated. These are often referred to as *tuned circular polarized* antennas, but due to their appearance, flat-pack just seemed appropriate, and a lot less cumbersome.

You can't find flat-pack antennas everywhere. Look for a large, well-stocked surveillance equipment store. Some of the companies listed in Appendix A, for example (such as Supercircuits or Polaris Industries), will have them.

These antennas are seen often in combination with high-gain receiver systems, but you will also see them on some of the small wireless video links. In the latter application, usually both the receiver and transmitter have a small (two inches square or so) radiator arranged in a flexible mount that enables you to position each antenna toward the other unit. The small size doesn't produce as much gain as the larger antennas, but it is a considerable improvement over the omni-directional whips.

The flat-packs vary in size, again depending on frequency, from around six inches square to as much as eighteen inches square, and most are between one and one and a half inches deep. They do display directional characteristics, but not nearly as prominent as the Yagi or dish devices. And, last but certainly not least, their size and shape lends itself well to covert placement. As long as you can aim the flat side in the general direction of the transmitter, the antenna can be behind a wall, in a closet, or in a variety of other less noticeable locations.

On a final note, the trusty old dipole antenna can also be used, to good results, with many of these systems. To refresh your memory, a dipole is simply two lengths of wire, cut to a half-wave length for a particular frequency, and con-nected with coaxial cable at the center. Simple, but still one of the best antenna designs ever conceived. Dipoles do not exhibit any gain, but their efficiency goes a long way toward not needing gain. They are easily home brewed and practically negligible in cost. If all else fails, give one a try.

Long-Range Wireless Sytems

I have discussed this before, but it does bear mentioning again. In terms of distant signal carry, it is amazing what some manufacturers have done with the Federal Communications Commission-approved microwave frequencies. The 1.2 giga-hertz band, 2.4 gigahertz band, and to some extent, the 900 megahertz band have all opened up new possibilities when it comes to wireless communications links.

These frequencies are commonly used for cordless telephones and computer in-house or local area networks (LANs), and now have invaded the video realm. The invasion is a welcome one, as the results are nothing short of stunning. Picture and audio clarity are remarkable compared to older transmission methods.

Originally, these links were short distance in nature, at least with video, usually 100 to 300 feet. But recent research and development has come up with some astounding equipment that can send the signals thousands of feet, sometimes even miles, and stay within FCC parameters for low-power devices.

One remarkably small 2.4-gigahertz system conservatively claims a three-mile range and states that up to ten miles have been achieved under line-of-sight conditions. It must be remembered that signals at these frequencies will travel almost unimpeded for great distances. They just don't bend or bounce off atmospheric layers very well.

The term *impedance* in this context refers to large, dense structures such as buildings, wooded areas, metal bridges, large electrical power stations, and so forth. The atmospheric layers will not cause a problem in terms of the signal passing through them; it's just that these frequencies don't bounce back to Earth. They cut through the layers and go out into space.

Hence, the microwave region has opened up a whole new ballpark when it comes to long-distance transmissions. Since a live video signal must have six megahertz of bandwidth, the efficient carry property of gigahertz transmissions does come in handy. After all, the microwave bands are hundreds of megahertz wide, so there is plenty of space for video channels.

Other systems boast 2500 to 5000-foot ranges using specialized receivers and antennas. Here again, design engineers make use of modern technology to improve the carry of video signals. The performance of ultrahigh frequency transistors and other semiconductors has come a long way in just the last decade and that accounts for much of the success manufacturers have enjoyed in extending signal range. There is little, if any, sign of a slowdown in this arena.

However, one of the most pleasing aspects of all this is cost. The three-mile video link I spoke of rings in at under $600. Five years ago that would have been an unheard of price tag. Such equipment would have started in the several thousand-dollar range. Along these same lines, the hot receiver/antenna combinations, boasting a one-mile line-of-sight range, can be had for well under $300. You could not have purchased even the antenna for that just a few years back.

If extended signal carry is required with your system, there is some impressive gear available to meet your needs. Considering the size and power requirements of this equipment, and the fact that it won't break the bank, the industry has made great strides in satisfying the requirements of a public fascinated with video surveillance.

COMPUTER NETWORK SYSTEMS

Where would we be today without the personal computer? They seem to be involved in just about everything we do these days, so why not home and business surveillance? There is a role they play here, but it may not be as prominent as you might expect. One of the things a computer does best is document information, and that would include video information as well. However, to store video the computer has to use vast amounts of its memory resources. Often, the advantages of employing your computer for surveillance simply don't justify the cost of expanding the memory to the necessary level.

However, storage is not the only way your PC might be helpful in a surveillance system. It can also be employed to transfer and monitor live video information, and that doesn't require much memory, as long as you are not trying to also archive that information. This section provides a brief, but hopefully educational, look a how your computer might assist you in your surveillance efforts.

Using the Internet

The Internet can be used to transfer the view your video camera captures to any place in the world that has Internet service. How is this done? Video can be rendered down to the same collection of 0s and 1s that the digital format is made up of. It just takes a lot more 0s and 1s than text or other data. As long as data can be made digital, though, the information can be sent over the Internet.

The pictures you see won't exactly be what you would see on a standard monitor. The transfer rate has improved over recent years, but is still slightly less than the thirty frames a second we are used to, so it has a rather jerky feel. If you can live with that, your computer may come in very handy.

Naturally, you are going to have to add some software to your PC. As always, the advantages have to be weighed against the cost. Also, such a system will tend to tie up your computer from time to time, and that might not be acceptable. The software is getting better all the time, and down the line, perhaps, all of this will work at lightning speed. For the time being, it is not the most efficient option available, but you should be aware that it is out there.

Worldwide Access to Your Surveillance System

One huge advantage of a computer-based monitoring arrangement is that surveillance can be conducted on such a large scale. As I just pointed out, the Internet provides one vehicle by which an observation system can be viewed at remote locations. The versatility this furnishes is more than obvious.

Another option is to set up privately controlled networks to do the same thing in a dedicated fashion. Now, this is not an inexpensive venture, mind you. It utilizes telephone lines, but you have to have the appropriate hardware and software installed at each monitoring locale.

I am alluding here to a dedicated system that requires both extensive planning and substantial revenue to create. Additionally, it is going to be one heck of a challenge to get up and running. But, for those of you who are not faint at heart, this might be the solution to an abnormal requirement. At least, it is something to keep in the proverbial "back of your mind" in case you need it.

Depending on the system, this may not be substantially different from setting up any other networked computer environment. If your big high-tech company already has a computer network, the security system might be built in. For a lot of people, setting up a computer network is much less scary than soldering wires on a PCB. For a company, there should be a person designated as, oh, a "computer workgroup manager" or the like, who should know whether the company system will handle surveillance. For do-it-yourselfers, you should be able to use the Internet.

ACCESSORIES FOR VIDEO SURVEILLANCE SYSTEMS

There is a whole truckload of stuff that could fit into this category, much of which we have already discussed to one degree or another. In this section I will dwell primarily on the accessories that will make your life as a surveillance installer easier. These fall into three areas: lighting, lenses, and camera supports.

Lighting

We have talked a lot about lighting thus far, so I will try not to be overly redundant in this section. As previously stated, CCD and CMOS cameras are sensitive to infrared (IR) light in the 800 to 1100 nanometer range, and this opens several doors when it comes to lighting.

Infrared wavelengths can be obtained from both solid-state light emitting diodes (LEDs) and more conventional sources such as incandescent bulbs and xenon flash systems (strobes). The latter require optical filtering to remove the visible light and usually don't get all of it (a dim red glow can often be observed from such devices). When that slight glow is not problematic, strobes can be a viable and strong source of IR light.

The light emitting diodes, on the other hand, will produce fairly intense infrared illumination with no visible signs. These small semiconductor devices can emit a variety of wavelengths and also come in the laser version. Diode lasers will provide a very high-intensity beam of IR light, but I think we have already discussed that in the lighting section.

As a reminder, commercial LED lamps and spotlights are available from a number of sources. These tend to be a little pricey for my taste, but if you don't want to take the time or don't feel you have the skill to construct your own IR light system, they may well be a welcome alternative.

Lenses

Here, you can really knock yourself out! Not to mention your pocketbook! Accessory lenses are available for any of the cameras that come with a C or CS mount, and the selection is nothing short of awesome.

Whatever your visual need, it is highly likely one or more of the video distributors will have a lens to meet that need. The variety is seemingly endless. Lenses come in wide angle, normal view, telephoto, pinhole, and zoom. They can have a fixed or adjustable iris or focus. They can be manual or motorized in terms of focus and zoom. The list goes on and on (see Figure 8.11).

One company offers a viewfinder that is helpful in determining which lens you will need in each camera location. Here, you look through the viewfinder and adjust an outer ring until you see the view you need. The markings on that ring will tell you what size lens to purchase.

PHOTO COURTESY OF SUPERCIRCUITS, INC. USED BY PERMISSION.

FIGURE 8.11 An interchangeable zoom lens

When I say *size*, I am referring to the focal length of the lens. That is, the distance at which the lens focuses. For CCD and CMOS cameras, wide angle will be in the three to five-millimeter range, normal is about eight millimeters, and anything above that is telephoto. What this means is that the image that emerges from the back element of the lens, which is the one that would fall on the film in a conventional still camera or the CCD detector in a video camera, will be sharp when it is three millimeters away from the surface of the film or CCD.

As their names imply, each type of lens provides a certain view. Wide angle offers a mid to very wide vista, while telephoto lenses magnify the image to cover only a selected section of the overall view. Zooms, of course, can change their focal length to zoom in and out on points of interest.

The pinhole lenses are designed for covert operation (see Figure 8.12). They only require a very small hole, often 1/8 to 1/32 of an inch in diameter, to see through. This allows cameras with pinhole lenses to be placed behind walls, paintings, in clothing such as in jacket pockets, hats or neck ties, or any number of other secret locations where a large hole needed for a more conventional lens would be conspicuous.

Some other interesting devices available include IR lenses, 2X converters, and adapter rings. The IR lenses are simply standard four, six, or eight-millimeter

PHOTO COURTESY OF SUPERCIRCUITS, INC. USED BY PERMISSION.

FIGURE 8.12 A pinhole lens

optics with a ring of infrared LEDs around the lens opening. This assembly furnishes both the viewing optic and the light source for near dark situations.

The 2X converters, or doublers as they are sometimes called, double the focal length of a lens. That is, if the lens is an eight-millimeter normal lens, with a doubler it becomes a sixteen-millimeter mid-telephoto lens. Neat, but of course, there are always compromises. In this case, the optical quality will usually suffer some. Not a lot, but you will normally be able to see a slight difference in sharpness. Doublers also soak up about half the existing light. Since most of the CCD/CMOS cameras are very sensitive to light, this is not usually a problem, but it's something you should bear in mind.

Adapter rings are used to convert C-mount lenses to CS mount. As you have probably already figured out, video camera lenses come with one of two standard mounts. Both are threaded style arrangements and one, the C, focuses slightly farther away than the other, the CS. Hence, it is necessary to convert the newer CS lenses to the focal length of the older C mount (see Figure 8.13).

Camera Mounts/Supports and Housings

It's important to understand that *camera mounts* or *supports* are not synonymous with *camera housings*. Mounting a housing will depend on what you want to attach it to, where you're going to place it, and a number of other factors too complicated to get into here. How you attach a camera to your house or business is really going to be up to you, but I'll cover at least the basic support units used in most situations.

Camera Supports

The next three figures show three views of typical camera mounts. Figure 8.14 shows a typical camera mounting unit that can be attached to the inside of a housing or to a wall, ceiling, or other interior area. The camera will have a matching threaded socket on the bottom or top or both, to mate with the male threaded bolt at the end of the mount extension. The holes in the large round section are for attachment to the surface (wall, ceiling, etc.).

PHOTO COURTESY OF SUPERCIRCUITS, INC. USED BY PERMISSION.

FIGURE 8.13 Permanent board camera lenses

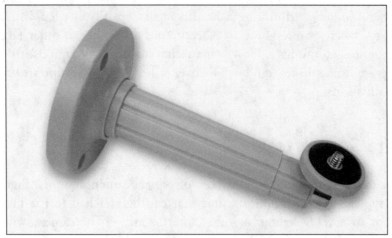

PHOTO COURTESY OF SUPERCIRCUITS, INC. USED BY PERMISSION.

FIGURE 8.14 Camera mounting unit for interior use

PHOTO COURTESY OF SUPERCIRCUITS, INC. USED BY PERMISSION.

FIGURE 8.15   Universal camera mount

Figure 8.15 shows a small universal camera mount with an extension rod. The rod can be used either in between the ball joint and the mounting plate or on the end of the male bolt that attaches to the camera.

Figure 8.16 shows a clamp-on-style mounting arrangement that also employs a gooseneck section for camera positioning after the clamp has been placed. These are normally used for temporary placement of a camera. Note, this shot shows a standard in-case video camera attached to the mounting assembly.

Camera Housings

I have discussed some of this before, so I will do my best not to bore you with a repeat of what has already been said. The primary purpose of revisiting this subject is to review the options when it comes time to actually locate the cameras. Since they are the heart of any video surveillance system, proper positioning is a must.

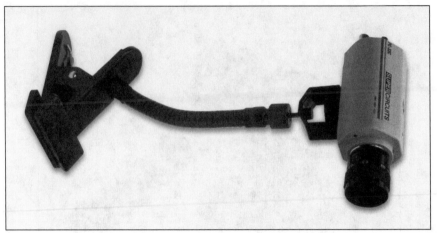

PHOTO COURTESY OF SUPERCIRCUITS, INC. USED BY PERMISSION.

FIGURE 8.16 Clamp-on mounting arrangement

In review, there are two scenarios regarding the location of your imaging devices (cameras). The first is indoor positions, and the second is outdoor. Each location requires certain considerations concerning the placement of the cameras.

Obviously, indoor locations don't necessitate protection from the weather. Pedestal-style camera supports are probably your best bet. These are those small devices that have a bracket at one end and an adjustable mount on the other that fits the standard 1/4 - 20 threads on the camera housing. They allow for just about any camera angle and are cheap.

If you will need protection against vandalism or other possible damage to the camera, then one of the housings will probably be a better choice. Here we have ceiling cases, corner housings, domes, and smaller protective cases for board cameras. Each type serves a special purpose in camera positioning, and I think the names describe that purpose. Naturally, these are going to cost you more.

For outdoor locales, weather protection is necessary. This is where the outdoor camera housings come into play. The housings made for exterior locations are fundamentally designed to safeguard cameras against moisture. These units generally incorporate watertight seals and gaskets for wiring in their quest to save your precious CCD imager from the perils of outdoor existence. The housings do a good job against wet weather.

Another type of outdoor housing not only keeps the wet stuff out, it also heats the interior to guard against very cold weather. Most of the solid-state cameras will operate at temperatures down in the near-freezing range, but in some locales, it is a good idea to climatize the camera's environment.

Between the two types, you should have all the protection you will need when it comes to weather.

Pan-and-Tilt Devices

For outdoor surveillance, few accessories beat a pan/tilt assembly (see Figure 8.17). That is not to say that you will never use one indoors, but you will find these devices extremely helpful when it comes to covering large exterior areas.

Applications might include parking lots, extended walkways, your front yard, your backyard, or any other wide or long area that would otherwise require multiple cameras for proper surveillance. The inclusion of pan (movement from side to side) and tilt (up and down) functions will allow a single camera to cover an extended viewing range and since many of these units are remotely controlled, you will be able to direct the camera's vista to where it is needed.

As with all video equipment, there's a wide variety of pan/tilt units to choose from. The money you want to spend is probably the primary factor regarding what you purchase, but versatility should also play a role in your decision. Prices range, as of this writing, from under a hundred dollars for a bare-bones unit to over $2000 for a complete computer-controlled, multifunction system.

Panning will vary from about ninety degrees from left to right to almost a full 360-degree circle, again depending on the coins you want to shell out. Tilt will range from twenty to fifty degrees, which may not sound like a lot, but once you have seen even twenty degrees of vertical change, it becomes clear how effective that is.

PHOTO COURTESY OF SUPERCIRCUITS, INC. USED BY PERMISSION.

FIGURE 8.17 Pan/tilt unit.

Of course, most of the systems have an automatic continuous pan default that will keep the camera constantly scanning its panorama. Often, that viewing range is set using the built-in stops on the base of the pan/tilt camera mount. These stops, which are usually adjustable posts, mark the start and end boundaries of the pan field and normally have to be set manually. There may be high-end systems that allow the range to be changed from a remote location, but if they exist, I have not found them.

Pan/tilt systems generally come as two units. The camera mounting unit contains the motors and gear assemblies for tilting and panning, although some of the basic units do not have motorized tilt. This section also contains the adjustable stops that set the pan limits. Panning speed usually ranges from about five to ten degrees per second, but some units can increase that rate. Tilt is normally around three to five degrees per second.

The second part of the system is the controller, and these can range from basic units that merely provide manual override to very complex devices that do everything but make coffee. With the advanced gear, computer control can determine pan and tilt speeds, camera settings, zoom and focus the lens, and even talk to another computer if that is necessary.

One system can actually digitally analyze pixel patterns in a subject and lock on to that subject. This way, the camera can follow the subject's movement. It also zooms and focuses the lens to keep the subject size uniform and the image clear.

This last system is often used for video conferencing and remote educational purposes. But if you want to spend the money, it could be a dandy home or business pan/tilt/zoom/focus/follow-the-action addition to your surveillance.

As might be expected, many of the controllers also have motion activation. With this feature, someone or something entering the camera's view initiates various device functions. It might also be used to start a videocassette recorder or other image preservation apparatus. Of course, the more elaborate controllers allow you to pick and choose what functions you want activated.

Standard Telephone Line Links

Applications of surveillance systems that utilize standard telephone wiring are becoming more and more popular. The only real drawback, as of this writing, is expense. These units come in a variety of configurations, but probably the two most prevalent versions are the single-service devices and the multitasking systems. Each has its purpose and each has its own price tag.

The single-service units are sold as a pair of transceivers that can be hooked up for a sole dedicated job. For example, you could set up a video/audio link between your electronics workroom and the kitchen, so that your spouse can remind you to come in and eat once in a while. In a business setting, you could install a link between the sales office and the shipping dock to ensure prompt delivery of your product.

This variety of the phone line video transceiver generally has a range of about 2000 feet for black-and-white signals and 1000 feet for color images. The really nice part is that they use existing telephone wiring and do not interfere with the phone system. They will not work with a private business exchange (PBX) installation, however. Also, you cannot send the signals out onto the regular telephone system. These devices are strictly for intraoffice or home operation.

These units are limited to offices inside one building (unless your company has dedicated phone lines between buildings or the buildings are extremely close together). You will have to have dedicated lines no more than 2000 feet long for black and white and 1000 feet for color.

Another nice part is the price. You will have to have a pair of transceivers for each individual link (camera), but they ring in at around seventy to ninety dollars per pair. Even a multi-link system would not break the bank. *Multi-link* is just a term used to indicate that you can use more than one transceiver pair. How many used would depend on the application needs of the person installing this type of equipment. They could be rigged through some form of controller, but I haven't seen anything commercially available along those lines—a controller would have to be home-brewed. Again, you could hook up as many cameras as you need, but each one will require a separate link. Also, the units incorporate high interference and noise immunity, so standard eight-conductor telephone cable, or twisted pair, is all that is necessary for clear reception and transmission.

The two applications mentioned above are just examples; there are many ways to use this kind of system. Monitoring parking lots, interoffice conferencing, and general building security are ways these transceivers can help your business operation. And as for the home, monitoring any room in the house or outside area are some of the tasks this system can manage.

All in all, if you are looking for a very portable link system that goes anywhere the telephone lines go, this might well be your ticket. To some degree, range and application are limited, but the overall cost is quite attractive.

This system is mostly for communicating, but you can set it up to just watch an area—in other words, one end of the link is on continual broadcast and the other end on continual receive. Just don't include a monitor on the camera end, or a camera on the monitor end.

If, on the other hand, you need a very complex and versatile system, a multi-tasking network is going to be in order. Be advised up front that you will spend a bundle on this one—in the range of $2000 to $4000. It is going to do just about anything your little heart desires, though, when it comes to transmission and reception of video and audio over existing telephone circuits.

Another thing to keep in mind is that the cost of the basic unit does not include the necessary computer system to run the show. This needs to be a PC with Windows 95 or better and at least a 33.6 kbps modem. You may well be able to employ an existing computer, which should help relieve some of the monetary pain, but you will still need to set aside some substantial money to purchase the rest.

I'm not telling you all this to scare you away from such a system, just to make you aware of the cost factor. If it's what you need, such a system might well be worth the expense, as these phone links can really perform. So, without further delay, let me get into what they can and will do for you.

With a multitasking telephone link installed, you will be able to monitor a business or your home from virtually anywhere in the world that has telephone access. In this age of satellite telephone service, that is pretty much everywhere.

With a multitasking link, the computer to store the video could be right with the system or it can be a "server" style arrangement. It might be just like a computer network, with a server somewhere storing everything, and maybe even running the applications, and then cameras with the monitors in remote locations, probably with their own hard drives and computers attached. These systems are going to come with complete installation instructions, and they ought to for what they cost!

If you want to, you can record video to the computer's hard drive, but as I mentioned in an earlier discussion, that does take a large amount of memory space. Additionally, you can retrieve those recordings using time and location parameters, and that makes it easy to reference the video archive. When playing back the video, you have the options of freeze frame, single-step frame, or scan to help isolate exactly what you are looking for. For alarm settings, the system offers automatic dial-up to let you know there is a problem.

Hold on, it gets better. The network supports both multiple units and multiple users, which can make the system a company-wide monitoring operation. The usual stuff like time/date stamping, multisite monitoring, printouts, and image saving, that are seen with complex multi-camera surveillance systems, are all there. This is a very versatile arrangement.

The possible applications are boundless. Monitoring your home, factories, warehouses, remote property, and road traffic are just some prospects. Security at remote and unmanned sites, and management of multiple retail stores are a couple more. With the addition of a couple of accessory packages, (which means more money), you can add auto transmission and remote tilt/pan capabilities.

As you can see, these are truly remarkable systems. When the need justifies the cost, the prospective advantages added to your surveillance capability by this technology are beyond question. In the end, it may well prove to be worth every penny.

Digital Still Cameras

There is a distinction between digital still cameras and digital video cameras. The only similarity is in the image detection elements. Both use CMOS or CCD detectors but the still cameras are designed to capture individual single images as opposed to the series of still images that make up the motion of "live" video.

The retail versions of digital still cameras first appeared about ten years ago and were primarily associated with computer imaging and video conferencing. They were able to produce some very rough and jerky video motion. Since that time, things have changed.

Today, technology has taken us from that early start to a variety of formats in digital video. Naturally, the dedicated computer cameras are still around. These are the small cameras of various sizes, shape, colors, and cost that normally sit on top of the monitor. I still have my old Connectix ball-shaped imager from days gone by, and it still works like a champ.

There have been marked improvements over the initial players in this field. Horizontal resolution has been greatly enhanced and the frame and field times have increased to the point where almost seamless video can be produced, recorded, and sent (little of that original herky-jerky motion).

Additionally, progress on the software side of this coin has vastly refined the quality of these imaging devices. In fact, today's dedicated computer video cameras are more than qualified for a simple surveillance system. They do not produce the results you can get with NTSC composite cameras and monitors, but if you already have the equipment on hand, you might give it a try, at least as a starting point.

Another form of the digital still camera has become immensely popular in the last few years. I am of course talking about the handheld devices you carry around just like you do with photographic film cameras (see Figure 8.18). Like their cousins, these record snapshots for posterity.

Most of the digital still cameras use small memory cards called flash cards that come in various sizes. Usually ranging from around two megabytes to thirty-two megabytes of storage space, these memory cards can be popped in and out of the camera. When you get back to an appropriately equipped desktop computer, the information on the flash cards can be downloaded to the system and used as needed.

All but the very low-end models have incorporated small 1.8 to 2.2-inch liquid crystal displays (LCDs), usually on the back of the camera, as viewfinders. This permits users to line up the shot and instantly see the picture they have just taken. If it is not right, then that shot can be deleted and the operator can try again. There is one really bad defect with these LCD viewfinders. They tend to easily wash out under light conditions much brighter than indoor illumination. This can make viewing shots in bright sunlight almost impossible. You can usually shade the viewfinder with your hand, but it is awkward.

One company uses regular 1.44 megabyte floppy disks for data storage in its cameras. Naturally, this makes things very convenient. One problem with the flash cards is expense. They can be costly. Floppy disks are down to less than fifty cents apiece these days, so you can carry a whole pocketful of them. Floppies are also easy to change out.

PHOTO COURTESY OF SUPERCIRCUITS, INC. USED BY PERMISSION.

FIGURE 8.18 Digital still camera.

One of the early drawbacks to digital still cameras was resolution. Pictures I saw that had been taken with some of the first cameras were anything but acceptable. If you enlarged them too much above postage stamp size, they began to fall apart. The images became grainy and soft or fuzzy. That has changed, too.

A combination of improved imager quality and higher record resolution (increased pixel size) has put these cameras into a whole new realm. The really high-end systems are being used professionally by commercial photographers and photojournalists. The visual attributes approach those of photographic film. It goes without saying this trend will continue.

Let me provide some general resolution numbers here in respect to horizontal resolution. As a rule, 250 to 350 lines is low resolution, 350 to 450 lines is medium resolution, and anything above 450 lines is high resolution. The NTSC format has 525 lines of horizontal resolution (the number of lines your TV will produce), but the $10,000 cameras are going to produce as much as 700 to 750 lines, which are then compressed to the 525-line format. The end result is higher resolution (better picture sharpness).

For a standard home or business surveillance system, these cameras are not going to be all that handy. But they are a practical and fun way to take pictures.

The last type of digital still system I want to touch on is the dedicated surveillance camera. These are rather sophisticated devices, with a price tag to match, which can be located in areas that are otherwise hard to monitor. The manufacturers suggest such things as catching vending machine thieves, identifying vandals, and monitoring drug drops as possible applications for their product.

Most of these cameras have motion activation and come in weather-resistant cases for outdoor deployment. The sensitivity of the motion detector can be changed and a date/time stamp can be included on each frame. Often, they also incorporate methods of activation/deactivation and checking system condition (number of frames taken) through magnetic sensing. This allows the operator to maintain the system without having to actually handle the camera.

So for applications involving individual frame recording, such as law enforcement observation, these cameras fit the bill. You may find a need for such a setup in your surveillance.

I'm presenting this information more for educational value than anything else, as when it comes to home or business protection, still cameras play a limited role. For long-term casual (for lack of a better term) observation, say wildlife studies and the like, periodic still shots have value. Or for a trigger system that could catch the intruder entering your property, they are beneficial. But for the day-in, day-out surveillance applications we have discussed in this text, still video images don't really make the grade. It is nice to be aware of what is available just in case.

CONCLUSION

If you're setting up a video surveillance operation, the majority of your focus will be on the equipment to capture and display images. As I've explained, today's market affords quite a variety of equipment, in terms of both power and affordability. What you choose to use for your initial setup and what you later add to the system depends largely on the extent of your surveillance needs and your bank account.

9

Special-Purpose Cameras

Chapter 8 provides extensive coverage of the basic camera equipment you'll need to set up any sort of general video surveillance. Those units are appropriate for almost any setup in which the camera can be—or perhaps should be—visible. For some situations, however, you need something a bit more discreet or something that offers more flexibility. This chapter focuses on these special-use devices.

> **NOTE**
>
> A word about terminology before jumping into this topic. You'll probably notice that this chapter frequently uses the terms *standard* and *professional* to describe the quality of the device. In those cases I'm talking about the image that the camera can produce. Many of the more esoteric devices can produce images of surprisingly good resolution considering their tiny size or unusual disguises.

DISGUISED CAMERAS

If you are like me and have a fascination with gadgets, this section should delight you. Here we cover the multitude of ingenious masquerades various manufacturers have come up with to hide the presence of small video cameras. I still find myself totally amazed at some of this stuff.

Naturally, most of these configurations are designed for covert application. As we go through them, you will see what I mean. Considering the variety of ways designers have come up with to hide the cameras, I can't think of very many situations where you would not be able to find a unit that would blend in with the surroundings at least well enough to do the job.

On that note, let me introduce you to the world of disguised video cameras.

Pager Camera

In this instance, a small board camera is concealed inside a pager shell. These come in monochrome and color, and can be worn wherever a normal pager would be worn. In this day and age of so many pagers, these cameras provide an excellent means of surveying events around you. The biggest drawback will be the angle of view, which is usually from about waist level. Pager cameras also come in a wireless version (see Figure 9.1).

Tie Camera

Here we have a small board camera hidden on the backside of a reasonably nice looking, conservative necktie (no hula dancers, so you won't look like a cheap gumshoe). Normally, a pinhole-style camera is used so that the opening for viewing can be as small as possible, and the perspective is great. Tie cameras come in both black-and-white and color, as well as standard and professional resolution.

Hat Camera

This is a regular baseball cap with a popular sportswear company logo on it, and with either a monochrome or color camera concealed in the front crown. Again,

PHOTO COURTESY OF SUPERCIRCUITS, INC. USED BY PERMISSION.

FIGURE 9.1   Pager camera.

this provides an excellent perspective and comes either standard or professional-grade resolution.

Eyeglass Camera

I know, this one is hard to believe. Here, the camera lens is placed in the bridge of the eyeglasses and fed via fiber optics to a small box that contains the rest of the cameras (see Figure 9.2). The box looks much like a pager, so could masquerade as such. The camera's viewing angle is going to be virtually identical to that of the person wearing the glasses. This one is a color camera.

Pen Camera

Here is another one that is nothing short of astounding. In this case, the camera is hidden in what looks like a standard ballpoint pen (see Figure 9.3). And the pen actually writes. I have only seen this one as a monochrome unit, but knowing this industry, one will probably be available in full color and wireless sometime next week.

PHOTO COURTESY OF SUPERCIRCUITS, INC. USED BY PERMISSION.

FIGURE 9.2 Eyeglass camera.

PHOTO COURTESY OF SUPERCIRCUITS, INC. USED BY PERMISSION.

FIGURE 9.3 Ballpoint pen camera.

Eyeglass Case Camera

This one comes in high grade and professional quality resolution, and is a camera hidden in a hardshell case designed to hold and protect eyeglasses (see Figure 9.4). The possibilities are limitless in terms of where this case can be worn

PHOTO COURTESY OF SUPERCIRCUITS, INC. USED BY PERMISSION.

FIGURE 9.4 Eyeglass case camera.

or laid down. The camera looks out from the clip side of the case, and perspective might be a little weird at times, but hey, you can't have everything.

Bag Camera

Coming in high grade and professional resolutions, and in color, this is a camera concealed in a standard shoulder bag. This one would be great for also hiding a transmitter and battery pack or a miniature VCR.

Jacket Camera

Here, either a monochrome or color camera is concealed in a standard denim jacket (see Figure 9.5). The camera looks out from the button of the left chest pocket, so the viewing angle is great, and who would expect a video camera to be hidden here? Also, the jacket has plenty of room for accessories such as transmitters or a VCR.

PHOTO COURTESY OF SUPERCIRCUITS, INC. USED BY PERMISSION.

FIGURE 9.5 Denim jacket camera.

Button Camera

Talk about clever! Here we have a camera hidden behind what appears to be a standard shirt button. The button is placed through a buttonhole in a shirt, and the rest is concealed by the shirt, cables and all. This one is color.

Videotape Camera

With this device, either a color or monochrome camera is hidden inside what appears to be a standard videotape cassette. The camera looks out from the spine of the tape, which has a title and other information on it. The cassette can be laid on a table or desk, filed on a shelf, or placed in any number of other positions that facilitate the surveillance you want.

Picture Camera

As you might expect, this is a camera concealed behind a wall picture or painting (see Figure 9.6). These come in color or black-and-white (the camera—the pictures are full color), and cable or wireless. Excellent way to hide a covert video camera in a room or office. The camera comes already mounted behind a painting, but can be removed if you want to mount it on a picture of your own. You would have to find a spot where the scene is at least a little noisy to disguise the lens. This is a pinhole lens that only requires a 1/8th-inch hole, however.

Overhead Sprinkler Camera

Are you ready for this one? For a business environment, it is hard to beat. Here, a color or monochrome camera is hidden inside what appears to be a standard

PHOTO COURTESY OF SUPERCIRCUITS, INC. USED BY PERMISSION.

FIGURE 9.6 Picture camera.

fire protection sprinkler head. The head doesn't work, of course, but the high angle and unlikely location make this a hard video device to detect.

Speaker Camera

This one utilizes a decorative audio speaker grill to hide the camera. It does also function as a standard speaker. This is the type of speaker often seen in false ceilings in office complexes. Again, the high angle makes this a natural for really good surveillance of a room or office. So far, I have only seen this one in monochrome.

Air Purifier Camera

Here we have a fully functional air purifier that has a color or monochrome camera hidden inside. Additionally, there is a 2.4-gigahertz transmitter in there too, so this one is wireless. It comes with a matching 2.4-gigahertz receiver and has a range of about 300 feet.

Covert USB Computer Speaker Camera

This digital camera is concealed in standard functional computer speakers that plug directly into a computer's USB port. In this fashion, the signal can be recorded on the computer's hard drive or sent out over a network arrangement. This one is great for catching office pilferers. Another nice aspect is that the camera gets its power from the computer.

Covert Pencil Sharpener Camera

Here, a wired or wireless camera system, either black-and-white or color, is hidden inside a standard functional electric pencil sharpener. The wireless version uses 2.4 gigahertz and has a range of up to 300 feet.

PHOTO COURTESY OF SUPERCIRCUITS, INC. USED BY PERMISSION.

FIGURE 9.7 Exit sign camera.

Covert Wireless Exit Sign Camera

Here a wireless system is concealed in a regular working Exit sign like you see in so many businesses (see Figure 9.7). The camera can be color or monochrome, and the transmitter uses 2.4 gigahertz for a clear, interference-free picture. This one also comes in a wired version.

Wireless Covert Emergency Light Camera

This is a color or black-and-white camera and transmitter in a working emergency light assembly. Like the Exit sign setup, this one offers 2.4-gigahertz technology and color or monochrome camera.

Wireless Covert Smoke Detector Camera

If they can hide cameras and transmitters in all this other stuff, why not put one in a smoke detector? A working smoke detector, of course. You can also get a wired version, and the cameras come in color and black-and-white. Naturally, the transmitter is 2.4 gigahertz.

Wireless Covert Wall Clock Camera

Remember the large, round wall clock that hung in just about every classroom back when you were in school? Well, that's what this is. Complete with 2.4-gigahertz technology, or wired if you prefer, this can be color or black-and-white. Naturally, the clock is fully functional.

Wireless Covert Portable Radio Camera

Here we have a color or black-and-white camera concealed in a working portable AM/FM radio. Complete with 2.4-gigahertz transmitter, this one could be placed just about anywhere surveillance is needed. The portability here is terrific.

Wireless Covert Clock Radio Camera

Concealed inside a working clock radio is a 2.4-gigahertz transmitter and video camera. Either monochrome or color, this one can also be ordered as a wired unit. Great for office surveillance.

Floodlight Camera

Here we have a little twist. This one is wired but it uses the same 120 VAC household line that powers the unit to transmit its signal. This is done through a proprietary induction RF system that rides the camera signal on the power line. Black-and-white only, range is approximately 2000 feet, and of course, the floodlight works. See Figure 9.8.

Wireless Covert Desk Lamp Camera

A standard working desk lamp hosts this wireless system. In either color or black-and-white, the transmitter uses the 2.4-gigahertz frequency with a normal range

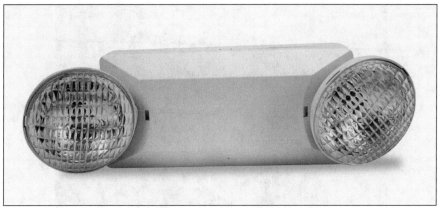

PHOTO COURTESY OF SUPERCIRCUITS, INC. USED BY PERMISSION.

FIGURE 9.8 Floodlight camera.

of up to 300 feet. Also, this transmitter allows you to select up to four different channels providing an interference-free signal.

Covert Convex Mirror Camera

This arrangement combines the standard convex mirror seen often in retail outlets with a hidden black-and-white camera. The camera is mounted behind the one-way mirror (see Figure 9.9). This is kind of a double-duty system, as you get both the observation mirror as well as the covert camera. The mirror does its job, providing the super wide-angle view it is designed for, while the camera gives a commanding sixty to seventy degrees of coverage. An interesting and innovative approach to area monitoring.

CAMERAS FOR SPECIAL SITUATIONS

Now, I will touch on some even more unusual packaging. If you have a special need, there is probably an existing system to meet it. In addition to the more commonly seen housings, such as the domes, corner enclosures, and surface cases

PHOTO COURTESY OF SUPERCIRCUITS, INC. USED BY PERMISSION.

FIGURE 9.9 Observation mirror camera.

associated with retail stores, as well as all the hidden camera gear I just covered, there are numerous special-purpose containers and arrangements. In this section, I will describe as many different examples of this as I can find. Some require special conditions for operation, such as a Federal Communications Commission permit for wireless equipment. But most of these devices are merely to satisfy a unique requirement.

Flexible Fiber-Optic Inspection Viewer

Here is a handy tool that will allow you to see in areas that are otherwise inaccessible. This apparatus, or something similar, has shown up in a number of Hollywood movies looking like high-tech equipment only available to the government. In reality, for less than $300 anyone can own this gear.

The secret to this device is the fiber-optic assembly that allows the image seen by a lens mounted on one end to travel to a video camera at the other end. A viewer is conveniently built into the receiving module (camera) that permits the image to be seen. That image will be anything in the vision of the lens at the other end. The fiber-optic cable, which can be shaped, may be positioned in such a way as to see behind or under objects. Additionally, and here is where Hollywood gets such a kick out of this gadget, you could drill a small hole, insert the fiber-optic assembly through the hole, and then use the device to see what is on the other side of walls, floors, or ceilings (see Figure 9.10).

All in all, flexible viewers have a great deal of potential for all sorts of difficult situations. They come with either eighteen or thirty-six-inch fiber-optic assemblies, have a built-in high-intensity lamp for illuminating the subject, and see in living color. If you have $300 and a need to see around corners or through walls, you might consider one of these gems.

Peephole Super-Wide-Angle Camera

Surely you have seen and used those little peepholes that extend through a door so you can see who is knocking on the other side. They are particularly prevalent in hotels and motels. Well, this is the electronic version of that device (see Figure 9.11).

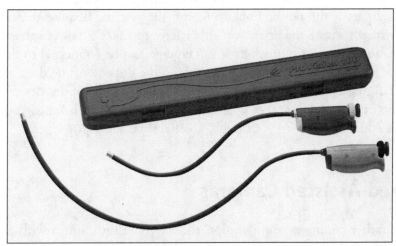

PHOTO COURTESY OF SUPERCIRCUITS, INC. USED BY PERMISSION.

FIGURE 9.10 Hollywood spy gear.

PHOTO COURTESY OF SUPERCIRCUITS, INC. USED BY PERMISSION.

FIGURE 9.11 Peephole camera.

It consists of a low light level (0.1 lux) monochrome camera with a 150-degree viewing angle on the front, which is on the outside of the door, and a small monitor on the backside, inside the door.

The thickness can be adjusted to accommodate just about any standard door. The only tricky part is the power hookup to run the system. Because the camera is located virtually inside the door, you either have to snake a power cable through the door's interior to a point where it can emerge and be connected to the power source, or figure a way to run the power along the outside of the door and keep the cable from getting in the way. Just like the optical version of this device, all that the person outside the door sees is an attractive stainless steel bezel. Even if this is not an essential addition to your door, it certainly is a clever gadget.

Infrared-Assisted Cameras

Another rather common configuration these days is the camera with a ring of infrared (IR) light emitting diodes (LEDs) circling the camera lens (see Figure 9.12). These cameras are sensitive to invisible infrared light and can use it to illuminate subjects in the dark. Not only does this provide a very practical light

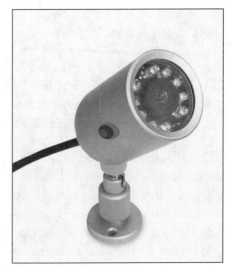

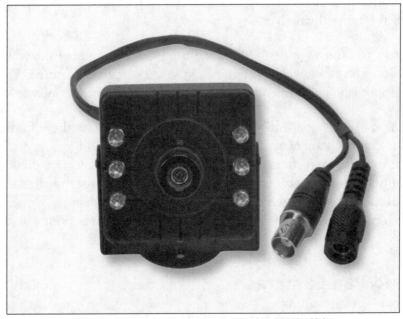

FIGURE 9.12 Infrared cameras.

source for covert installations, it also allows the camera to see at night in such applications as front-door monitoring, and child or sickroom observation.

One drawback to this setup is that the LEDs don't cast their light very far. Usually about ten to fifteen feet is all the farther that you can go and still get a perceivable image. In many situations, though, such as the ones mentioned earlier, that is more than enough to accomplish the goal.

These IR-assisted cameras come in a variety of shapes and sizes. All of the normal camera configurations, such as square boards and lipstick cameras, have an IR version. Some are small with just a few LEDs, while others employ as many as a dozen emitters. Although more LEDs do not really give you much more viewing distance, they do increase the illumination in the area, giving you a better view of the area and any intruders.

However, there is a distinct limit to just how much better the illumination can get, regardless of the number of LEDs. For a really long-range light source, IR laser diodes are the recommended form of emitter. If properly aligned and aimed, they can furnish illumination up to about 100 feet, depending on the milliwatt rating of the lasers themselves.

In addition to cameras with the light emitting diodes mounted directly on the boards, a number of individual infrared light sources are manufactured. These can be round spotlight arrangements or square pads, but all produce intense IR light.

Often the spotlight types utilize optics to better concentrate the infrared light, and some claim distances as far as 200 feet. I have not had an opportunity to experiment with any of these units, and unless they employ laser diodes, I am reluctant to believe those distance figures without seeing it for myself. Also, these light sources are pricey, often costing considerably more than the cameras. It is the old story, however. If you need the light, then the cost is probably justified.

Tilt-and-Pan Cameras

A really nice feature to have for surveillance cameras is the ability to pan over a larger area than a lens can view in a single frame, and also to be able to tilt the camera up and down when needed. I think it goes without saying that these abilities expand the versatility of the camera position. To accommodate this need, manufacturers have come up with mounting stations that provide both functions (see Figure 9.13). Be advised, though; the prices seem steep for what you are getting.

Some of these units are remote controlled; others are activated by motion, change in ambient light, or sound; and some can handle larger cameras or weatherproof outdoor camera housings. Just about all of them start in the $150 range and go up from there. Here again, though, the need may justify the cost.

These systems are really more valuable to businesses than home installations. It is nice to be able to pan over a parking lot or a large assembly facility for employee safety. For home application, they are good for a large front yard, backyard, or similar area that is wide in nature.

PHOTOS COURTESY OF SUPERCIRCUITS, INC. USED BY PERMISSION.

FIGURE 9.13 Tilt-and-pan camera.

FIGURE 9.14 Underwater camera.

The tilt function is again handy in situations where large spaces need to be surveyed. The parking lot example is a good one, as they are traditionally both wide and deep, and the tilt operation allows for full coverage.

Underwater Cameras

This is an area that can be both functional and just plain fun. There are several different versions of cameras that are made to go beneath the surface of a body of water. Some are just simple waterproof containers with a marine grade cable attached to the unit for raising and lowering. A power line for the camera and the video output cable are usually fastened to the main cable to keep the system as contained as possible (see Figure 9.14).

Another system I have seen advertised looks like a miniature submarine. The camera sits on top of two cylinders with a rudder connected at the rear. Heavy mesh cable is used to raise, lower, and retrieve the unit, and the power and video are carried over sealed electrical lines. This unit claims possible depths of 300 feet, and the camera has infrared LEDs as a light source.

The underwater cameras do come in both color and black-and-white versions, but due to the heavy blue overcast experienced at any depth at all, color might well be an unneeded expense for this application. You might well be better off with a more light-sensitive black-and-white model.

If you enjoy studying objects beneath water, whether it's a lake, stream, river, or ocean, than this is probably something you will want to investigate. These cameras are quite useful for such things as treasure hunting, search and rescue, sport fishing, and marine research. All in all, they should be of interest even to those who don't have a burning curiosity about what goes on underwater.

CONCLUSION

Whether your situation requires filming underwater, around a corner, or just in a less noticeable way, the special-purpose cameras discussed in this chapter can help you accomplish your objective—capturing the image.

10

CHAPTER

Special Surveillance Equipment

Manufacturers normally target some special areas when developing any product, and video surveillance gear is no exception. Here, however, the equipment can get quite specialized with a very limited market in mind.

What I am getting at is gear designed for law enforcement, the military, and the intelligence community. I bring this stuff up more for your educational gratification than anything else, as we are talking about stuff that you and I cannot purchase, unless you are involved with the FBI or CIA, or know some questionable people.

So let me conduct a quick tour of some of the more exotic equipment available. Available, that is, to a very select group of individuals and agencies. Enjoy the excursion!

LAW ENFORCEMENT

I will cover law enforcement first, but bear in mind that a lot of this gear overlaps into the other areas as well. For this equipment category, the military and intelligence people employ much of the same stuff law enforcement uses. There are some exceptions, and that will be covered as we get to each area.

As might be expected, law enforcement, which covers local and state police and federal government investigative organizations, makes good use of audio bugging equipment. Small wireless mics, as well as hard-line listening devices, are commonplace in the performance of these department's duties. Various companies that produce such equipment are more than happy to accommodate law enforcement.

Most of the gear, however, would be deemed illegal for a civilian to use, and many distributors of this kind of gear will not sell to anyone who does not have the proper credentials. That is probably good, as there are individuals, some of whom we have mentioned before, who would most certainly take advantage of this sophisticated equipment for their own personal gain. And I do mean at the expense of others.

With that said, we are talking about very small, very sensitive, long-range wireless transmitters that pick up audio only. Additionally, there are "tails" that can be attached to vehicles, and in some very special cases, people. These devices allow police to follow subjects from a distance without losing them. This is done by monitoring and plotting a radio signal emitted from a transmitter that the individual or vehicle is carrying, usually without the subject's knowledge.

The range of these devices can often be several miles, and that suggests a lot of power output. In most cases, that is exactly how the transmitters obtain such range, which makes them illegal to use under the Federal Communications Commission's (FCC) low-power regulations. This is another reason "average Joes" are not supposed to get their hands on this stuff.

Okay, enough about audio! Let's see what goodies are available for video! This is an area that law enforcement is taking a very serious look at. With the help of dash mounted surveillance and all the new super-small TV cameras around these days, video has really helped out in both arrests and convictions.

Naturally, the omnipresent dollar bill often dictates what type of gear a law enforcement operation uses. In scenarios where off-the-shelf equipment will do the trick, it only makes sense to use that gear. However, if more sophisticated stuff is needed, the manufacturers have it available.

Special police gear comes in a variety of shapes, sizes, and flavors. There is the very small gear meant to be worn on a person. There are lenses made that will see through walls, with the help of a small hole. There are also entire kits meant to handle just about any need a police department might encounter.

Again, special antennas, receivers, and normally illegal power levels all make this gear far more efficient than most of the "run-of-the-mill" cameras, transmitters, and the like available to us little folk. If you have a significant interest in this kind of gear, some of it will make your mouth water. Tiny powerful transmitters, micro video cameras, portable monitoring and recording equipment, and receivers that hear a flea-power signal at thirty paces are just some of the devices available to people with the right papers to back them up. If video is your bag, the special law enforcement gear will turn you into a kid in a candy store.

MILITARY EQUIPMENT

As said at the outset, what is used by law enforcement will also be employed by the military, at least in many instances. All the higher-power transmitters, micro cameras, clandestine lenses, and other stuff are used for military purposes, and the off-the-shelf gear comes in handy, too. But, just as for the intelligence community, there is some other stuff as well.

Due to the highly secret, hence highly classified, nature of these devices, not much is known outside the "need-to-know" circles. However, tales abound regarding voice transmitters the size of the head of a pin, super tiny video cameras, transmitters that can automatically change frequency or compress entire conversations into a split, second transmission, and more.

It is often hard to distinguish between reality and fable in this arena, as equipment descriptions are frequently embellished. With the release of much of the classified information from the former Soviet Union, though, we have at least gained a glimpse of what the USSR had to offer, and much of it is mighty impressive technology.

I think it is safe to assume (although no assumption is ever completely safe) that the United States military has comparable equipment. The microphotography that has brought us the astounding microprocessors used by our computers has also played an important role in the production of this tiny sized surveillance gear.

INTELLIGENCE COMMUNITY

These are the people that have not only benefited greatly by the miniaturization of transmitters and cameras, but are also very likely the folks responsible for much of the research that brought it about. In the space program, necessity was very much the mother of invention, and the same is true of the needs of our "spy agencies."

Like law enforcement's use of what would otherwise be illegal gear, the application of sophisticated spying devices by the likes of the Central Intelligence Agency (CIA), National Security Agency (NSA), and Naval Intelligence has made our world a safer place. During the Cold War era, for example, the nation was able to learn enough of what was in the mind of our enemies through the efforts of the CIA, NSA, and others, to keep us safe and free. Much of that knowledge was acquired through the use of less complex devices than we have today, even though they performed similar tasks. So this gear can not only keep your home and business more secure, it is helping protect the country that hosts your home and business.

In the way of illustration, here is a story: when the United States inspected its new embassy in Moscow, well over a hundred bugs were immediately detected and removed. Basically, those were the ones the Russians wanted us to find, hoping their discovery would satisfy the intelligence folks and the rest would not be found. It didn't, and most of the bugs were eventually discovered. When inspection teams x-rayed the walls of the building, some of the superstructure just "didn't look right." Instead of normal up and down studs in the walls, there were a number of round concentric circles found in several places. It was later realized that these were antennas designed to send out the information the audio bugs were picking up.

So the stories go. For its entire existence, the intelligence community has made the best use possible of electronic technology. Naturally, all these tiny transmitters and video cameras have had to be of tremendous assistance in the performance of their job.

CONCLUSION

Next time you hear or think about the CIA, stop to remember that this organization protects our country using some of the same equipment you use to help secure your home. The small board cameras that have become so popular, commonplace, and cheap are just as useful to the agency as they are to us.

Also, keep in mind the job law enforcement and the military does with this equipment. Again, much of the freedom, security, and peace of mind we enjoy today has come from the dedication of those people, often employing tiny TV cameras and other video surveillance gear.

PART FOUR

Special Issues

Before you start attaching wires or applying the screwdriver, read this section to learn about how to select the appropriate location for your equipment and how to maintain it once it's in place.

I've spent a lot of time to this point talking about how you can *observe* other *people*. Has it occurred to you yet that someone may be watching *you*? We've devoted a whole chapter to how to protect yourself from espionage.

Finally, this section winds up with a few pages on the future of surveillance. As electronic equipment of all types continues to shrink in size and price, and as a post–September 11 world grows ever more suspicious, the desire for more sophisticated surveillance continues to grow. This section addresses some of these issues.

Location Considerations

Obviously, when determining where to position video surveillance equipment you must consider angle of view. But there's more to consider than just the widest range of visible area. You must determine whether you want cameras to be visible or hidden, how to protect the equipment from vandalism or accident, how to set up the lighting, whether the equipment will require wiring (and if so, how best to run the wiring for safety and concealment) or will be wireless (and if so, how to supply power and ensure the best transmission and reception), and so on. This chapter explores all of these possibilities, with a special focus on remote setup considerations.

DETERMINING WHERE TO PLACE CAMERAS

Naturally, you want to position cameras where they provide views of the areas that need attention. But how do you go about this? There are several factors to be considered. For instance, do you want the cameras to be seen? Do you want a bird's-eye view or an eye-level view? Another important issue is protection for the camera. Do you want people to be able to easily get to the cameras and spray paint the lenses black? (Probably not.)

As for being able to see the cameras, there are two approaches: overt and covert placement.

Overt Placement

In the *overt* approach, cameras are in plain view. Persons being viewed will know they are being viewed and presumably be on their best behavior. Don't sell it short, especially in a business environment. Employees are far less likely to misbehave if they know there is a possibility that they are being taped.

I say "possibility," as many businesses have dome-camera arrangements where it is impossible to tell whether there actually is a camera in the dome or if it's just a decoy designed to keep employees and customers on their toes. On the average, the empty domes are about one-third the price of a dome with a camera (see Figure 11.1). This does, of course, depend on the dealer you buy from and the type of dome you choose. They are available with a variety of surfaces, such as reflective metal coating, smoked or clear plastic, and so on. Since the cost of some of these dome units, with cameras, has dropped to below a hundred dollars, in small areas you might want to have all domes armed. Where larger areas are involved, the empty domes can be an effective deterrent.

Whether the domes contain cameras or not, the overt location can be useful in deterring misconduct, illegal or bad behavior, or criminal activity or intent. The bottom line here is that the presence of a video camera, or the possible presence of a camera, will go a long way toward keeping scoundrels, crooks, thieves, bad guys, and so forth in line.

Covert Placement

The ability to watch an area secretly also has its advantages. Anytime you need to catch the culprit in the act, the covert approach is usually the better method.

For example, in a business situation, shoplifters cannot be prosecuted unless they actually leave the business with the stolen property. However, if they know they

PHOTO COURTESY OF SUPERCIRCUITS, INC. USED BY PERMISSION.

FIGURE 11.1 A dome camera assembly.

have been seen on a video camera taking the merchandise, they probably are not going to leave with it. This may well be an argument for the overt cameras, but there are some crafty crooks out there who have found ways around overt observation. Often, they have the ability to make off with the goods right in front of the camera. When dealing with these people, it is best they don't know that they are on camera.

For home protection, covert cameras are not highly recommended, at least for exterior or perimeter coverage. In this scenario, it is best for anyone with bad intentions to know he or she is being watched. It is far more likely that potential intruders will think twice and move on to the next dwelling if they see even a possibility of video surveillance.

This is akin to the lock adage. It is said that locks only keep the honest people out. But the more locks you have, the more reluctant a dishonest person is going to be about trying to get through all of them and into your home or business. Unless you have been "targeted," a criminal will more than likely just move on to another structure. Locks do serve a valuable purpose. The same goes for the very conspicuous video cameras.

> **NOTE**
>
> Of course, most people who have "nothing worth stealing" do not bother with video equipment. Loading up your home with lots of video equipment may suggest to the canny burglar that there are plenty of goodies in this location and may present a nearly irresistible challenge to some thieves. Consider a combination of overt and covert equipment, rather than having your property bristling with visible cameras that may invite intrusion.

Covert observation is also handy for watching an area you don't want anyone to know is being watched. This might include a child's room (as with the so-called *nannycams*) or sickroom in a home, a break room, returned merchandise department, or office in a business. If you either don't want anyone to know surveillance is going on or you don't want to upset the occupants of the area, hidden cameras are the best route to take. As discussed in the section on unusual camera configurations and disguises in Chapter 9, "Special Purpose Cameras," there is a multitude of ways for the camera to do its job and still stay out of sight.

In addition to protecting valuables and property, businesses should bear in mind the expectations of customers, clients, or patrons when setting up surveillance systems. In some cases, overt setups alone may be best. In others, a combination of overt and covert equipment may be the appropriate plan. For example, highly visible cameras in the parking lot, on building exteriors, and throughout building interiors is probably suitable in a daycare center—indeed, such placement may be reassuring to nervous parents. But less-noticeable equipment is probably more appropriate in a facility that deals with older children or adults who may be nervous about being filmed—for example, a skating rink or a grocery store. Businesses that deal with money, precious minerals or metals, or other valuables in any form should probably advertise the presence of extensive surveillance;

businesses that provide personal services (bed-and-breakfast hotel, marriage counseling, wedding services, and the like) must be more discreet.

Placement of individual cameras will depend largely on what you want each device to do. In many situations, a high angle is preferred, as it gives a broader view of the area under observation. Also, this placement will put the camera high in the room, where false ceilings and attic space can be utilized for power supplies and cable runs. Always consider a high corner location when evaluating your placement options.

If you use wireless systems, the cables will be eliminated, but you will still have to provide power for the camera/transmitter. Here again, an elevated locale provides access to what is above the room, making installation easier and quicker. Also, if maintenance or repair becomes necessary, the false ceiling or attic scenario makes that much less problematic.

For most applications involving home and business surveillance, clearly visible, "in-your-face" installations are far more effective. In the long run, being aware that they are on camera, and possibly being videotaped, goes a long way toward keeping people from doing something illegal or against the rules. To that end, let them know they are on TV.

Underwater Cameras

How is underwater surveillance of use? For home applications, monitoring activity in the pool is about the only use. However, for some businesses, it could be very handy. Any industry that incorporates submersion equipment or procedures could monitor those operations with underwater gear. That is, as long as the solutions in the tanks are not so corrosive that they eat your camera. Also, storage and condition of liquid components for manufacturing can be watched via closed circuit television. Again, don't subject an unprotected camera to anything that might damage it. (See the discussions later in this chapter on protecting your gear.)

Additionally, underwater surveillance might be pragmatic for amusement parks where water is part of the entertainment, in aquariums for keeping an eye on the

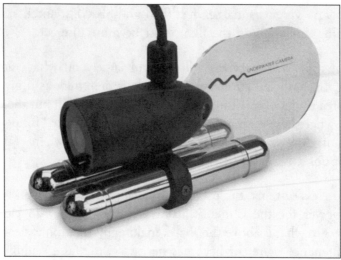

FIGURE 11.2 An underwater camera system.

"stars of the show," or virtually any other scenario where water and/or other liquids might prevent an unobstructed view. In such cases, surveillance for productivity and safety is a prudent notion (see Figure 11.2).

Camera Housings

Most distributors of video equipment carry an extensive line of camera housings made for this type of installation. These come in ceiling-mount, corner-mount, and universal-mount configurations and are made of either metal or plastic. The construction material usually depends on where the enclosure will be located— inside or outdoors.

Concerning which type of housing you may want to employ, a good rule is to use the sturdier, more expensive enclosures outside, while the less expensive units normally do well inside. The heavy-duty exterior housings can be made of either metal or plastic. You need to keep in mind that metal will conduct heat from the sun during daylight and cold at night. The plastic units are more than weather resistant if you buy one designed for outdoor use and endure extremes of temperature as well as metal without the conductivity. Most of the newer cameras have a reasonable tolerance for temperature change, but that can't be carried to extremes. For areas where extreme conditions are going to be encountered, exterior housings are made with internal temperature and humidity control systems, but be prepared to shell out some bucks for those.

In one of the surveillance systems I set up, I used homemade housing constructed from clear quarter-inch Plexiglas. The seams were sealed with clear silicon caulking material to make it watertight. These units worked great for a number of years and provided a very overt presence, which was quite appropriate for that particular surveillance setup.

For interior use, a number of very pragmatic designs are available. The corner units, that literally fit in the ceiling corner of a room, are great for overt arrangements, and, of course, the dome does a good job of housing and protecting inside cameras. Both dealers and catalogs that carry video surveillance gear will be able to help you decide which housing will work best for your situation.

A variety of pedestal-style brackets are available. These secure to ceilings, walls, or other sturdy surfaces and allow you to position the camera at just about any angle. They are less expensive than the actual housings and quite appropriate for many applications. Normally, your individual requirements will help determine what type of mounting apparatus to buy.

NOTE

If you're using a pan-and-tilt camera, be sure there is ample clearance in all directions for the camera/mount assembly. Repeatedly dragging a lens against a corner isn't particularly good for the equipment.

Naturally, we are talking overt placement here. With the aide of pinhole lenses, false sprinkler heads, smoke alarms, and the like, covert placement high in the room is also possible. In each case, the false ceiling or attic will facilitate positioning of the device, as well as maintaining its cover.

Regarding other locations not so high up, bookshelves, podiums, and even "on-desk" positions will work well for both covert and overt arrangements. Bear in mind, the lens coverage will be a factor with some of these spots, but trial and error will usually provide a good compromise or solution. Does the camera do what you want it to do? That is always the question to ask when setting up your surveillance system. If the answer is yes, then you have chosen wisely. If it is no, keep trying. You will find the right spot.

This discussion may seem unrelated to the original topic of this section—camera housings. But it's important to realize that in a covert system, everything comes into play, from cameras positioned inside walls or ceilings to cameras inside lamps, false books, or radios—and the host item in effect becomes the housing.

Temperature Issues (Climate Control)

So far, we have talked primarily about indoor locations. There are going to be areas outside your home or building you may want to watch, though. For a business, the parking lot and front entrance immediately come to mind. As for your home, the driveway, children's playground or pool area, as well as the front entrance, are natural locations for any surveillance system.

However, any outdoor installation does require a little care. The most important aspect to consider is camera exposure to the weather. As remarkable as the present-day solid-state cameras are, they are electronic and don't particularly like water, especially in the form of rain or high humidity.

When deciding on an exterior location, it is necessary to evaluate the amount of wet weather exposure the equipment is going to receive and choose a suitable housing. This is not as hard as it might sound. If you determine a camera is needed on the north corner of your house, monitor the next few rainfalls to see how wet it gets out there. From that information, you will be able to purchase an outdoor camera housing that will suitably protect your precious camera.

Incidentally, there are exterior housings that have built-in humidity control for use in environments that could be damaging to the delicate electronics of the camera. Do not overlook this aspect of the weather when appraising your outdoor positions. Prolonged exposure to high humidity can be just as bad, if not worse, than direct rainfall regarding the health of your equipment.

> **TIP**
>
> A simple and inexpensive trick for preventing humidity problems is simply a small-wattage incandescent bulb. Something in the seven to fifteen-watt range will do fine. Careful placement of the bulb inside the housing will help keep the inside atmosphere dry, or at least drier. I say careful placement, as you will want to keep the bulb at a reasonable distance from the camera itself. The principle is that the bulb will generate heat, which in turn will dry out the air. That is all well and good, but you do not want that same bulb baking your camera. Take heart, though, most outdoor housings have plenty of interior room to position the bulb at a safe distance from the camera. These housings are made for a variety of camera sizes and there is normally ample room at the rear of the housing when using today's smaller cameras.
>
> If your camera has an auto-iris, though, be careful using a light bulb. The extra light could fool the iris into thinking it is daylight and render your night video too dark. Another thing to consider is that light bulbs have a limited life span. Their average life is approximately 1000 hours, so if you use this method you must be proactive on maintaining the bulb.

Heated housings may be needed in areas where winter temperatures are severe. There is a limit to how cold (or hot) CCD and CMOS cameras can get before they simply stop working. While not always the case, such a stoppage may well do permanent damage to the camera.

So, if in doubt, it might be best to include temperature-controlled enclosures in your plans for exterior camera locations. Depending on the camera you buy and the dealer you purchase it from, a look at the manual or a quick telephone call might help determine the unit's susceptibility to extremely cold or extremely hot temperatures. Heat is generally not as much of a concern as cold, as the cameras themselves do generate some heat of their own and are made to withstand high temperatures. To avoid damage to the equipment and to protect the integrity of your surveillance system, this is one consideration well worth your time and effort.

Constructing a Custom Housing

Naturally, there is always the "home brew" approach to housings. It is not that hard to construct your own exterior housings if you want to take the time to do it. Quarter-inch clear Plexiglas is an excellent material to use for such a project. It is easy to work with and provides both a secure environment and visual access to the camera. Special glue made for PVC pipe is perfect for cementing the sides and ends together. This stuff is great. It literally melts the pieces of plastic, so when they are placed together and allowed to dry they become practically a single piece of plastic. Additionally, the thickness of the Plexiglas allows for small screws to be used along the edges. This reinforces the entire assembly. I recommend applying a silicon sealant to the inside seams of the box. That further protects the enclosure against moisture seepage.

Rubber seals can be employed at the rear of the case for cables and power connections. If you need a heating element, such as a small light bulb, the seal accommodates the wiring for this as well. On a final note, it is best to have one surface removable. Usually the front panel or top is a good selection. Using small screws and silicon sealant, you can maintain a watertight seal and still be able to gain access to the interior when necessary. This housing is weatherproof, and does not require any additional protection for the camera.

Fighting the Elements

Regardless of whether you're positioning surveillance equipment outdoors or in, you must consider Mother Nature. Electronic equipment is sensitive to extremes of temperature and humidity, as previously mentioned, but it also doesn't particularly like intrusion by dust, mold, mildew, insects, animal fangs, or the curious fingers of your children or your guests. Chapter 12, "Protecting and Maintaining Your Surveillance Equipment," addresses many of these issues in detail. Before you even purchase equipment, however, it's important to consider how to protect it. If you're going for exterior surveillance, for example, you need to be aware of animals in your area that may be attracted to equipment.

Following are just some of the "elemental" issues you should consider:

- **Power.** Gadgets that share circuits may be subject to temporary brownouts or circuit overloads. If possible, give your surveillance equipment its own dedicated circuits. If not, at least avoid placing cameras on the same circuits as computers, large stereo systems, heavy-duty printers or copiers, and other electronic equipment that requires a lot of power or can cause surges. If the equipment is battery operated, make sure that the battery is kept dry or corrosion may result. It's not just a matter of replacing the battery in that case; corroded terminals may eventually make the equipment inoperable.

- **Magnetic equipment.** Avoid placing cameras near magnetic sources such as vacuum cleaners, audiotape erasers, or refrigerator magnets, any of which can affect the quality of video reception.

- **Birds.** During mating and nesting times in particular, birds may be interested in anything shiny. Even one-piece or solid units can sustain heavy damage from a bird that attempts to wrest the camera from its housing. And don't discount the damage potential of ducks, geese, and other waterfowl as well as birds of prey. Birds with strong beaks or bills can break bones or nuts; imagine what they can do to thin plastic! Claws, talons—even webbed feet—can scratch. Finally, any piece of equipment that can provide a perch for a bird is subject to whatever it leaves behind.

Bird excrement is at minimum unsightly, it certainly can obscure a lens, and its chemical properties have the potential for serious damage if ignored.

■ **Bugs.** Roaches, beetles, flies, ladybugs, and those nasty Japanese beetles, just to name a few, love to cuddle up to a nice, cozy, warm microchip in the winter. If your camera's housing is not sealed properly, the floor of your camera could easily accumulate an inch or more of bug carcasses each season. This could ruin a clear dome. Even worse, it can destroy a camera. Cockroaches do conduct electricity. I've seen it.

■ **Beavers, raccoons, squirrels, bears, cats, monkeys, children....** Although this may sound like a diverse group, the issue is that anything with claws, opposable digits, actual fingers, and curiosity is a threat to equipment.

LIGHTING

Even as sensitive as some of the new black-and-white cameras are, they still need proper light to see. The term *proper* refers to the right type and amount of light necessary for the sensor to discern an image. With most CCD and CMOS cameras, the amount of light has become next to irrelevant. Let me qualify that statement. Nothing, including you and me, can see in total darkness, so there does have to be some illumination.

With lux ratings in the 0.1 and even 0.01 range, very little light is necessary. To refresh your memory, one *lux* is the amount of light produced by one candle at one foot. Imagine how miniscule one one-hundredth of a lux is. But that's enough for many of the presently available monochrome cameras.

Color cameras are trying hard to catch up with their black-and-white cousins; I've seen some advertised at one lux. Due to the nature of color imaging, though, color cameras require more light than black-and-white cameras.

The point here is that present-day equipment is very sensitive to light. When compared to even the highly sensitive vidicon tubes of yesteryear, these imagers are nothing short of miraculous in their ability to produce usable pictures from practically no light at all. The amount of illumination is usually the least of your concerns.

Light Source and Direction

What will be of more concern will be the light source. Does the source provide light from behind the camera or in front of it? This factor will make a huge difference in how well your surveillance system functions. In photography, a common rule is to keep the light behind you. It that fashion, your subject is suitably illuminated and will show up well in your picture. The same formula applies to a surveillance system. Fortunately, many modern cameras have *back light compensation* (BLC) circuitry that helps adjust for situations where the lighting source is in front of the lens, but there are a few things you can do to ensure the best possible image capture.

If you are installing lighting, it is best to position your light source close and slightly above the camera. In that fashion, adequate lighting is projected onto the subject the camera is viewing. Light sources from off-camera locations tend to create shadows that might make it harder to identify your intruder.

If you are relying on natural and ambient light, try to find locations where the lighting is from behind the camera's lens. Of course, this is not always possible, especially when dealing with sunlight. After all, it changes position throughout the day. But some judgment applied to the location of a camera can minimize the effects of the sun's changing location. Direct sunlight in the lens of your camera will not damage its CCD or CMOS image sensors, so that is one issue you do not need to worry about.

As for night viewing, floodlights make excellent illumination. They can easily be placed to accommodate both the camera's location and the area you want to light, and their cost is very attractive. They do produce rather harsh light, in terms of shadowing, but they will do the job when it comes to lighting your front lawn or driveway.

Infrared Light

CCD and CMOS cameras are sensitive to infrared (IR) light, normally in the 1000 nanometer range. That should conjure up all kinds of possibilities. This goes for both black-and-white and color cameras, although in a color camera, the infrared will throw the color balance toward the cold (blue) side.

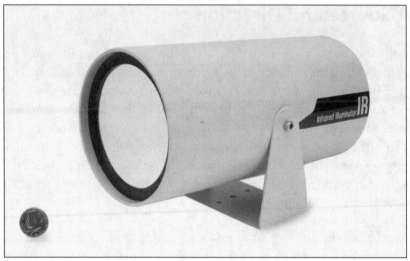

FIGURE 11.3 IR spotlight.

What this means to your surveillance system is a source of invisible light that your trusty little cameras can see. With the availability of high-power infrared light-emitting diodes (LEDs), some interesting lighting sources can be developed. Generally, ten to fifteen feet is about all that can be expected from the LEDs but often that will be more than enough. For instance, you might want to be able to see who is banging on your front door in the middle of the night without them knowing you are watching. Infrared LEDs would work great for that application.

In case you need more distance, several companies offer spotlight-style IR illumination sources (see Figure 11.3), or you can even make your own. In the same way that visible light can be concentrated with reflectors and/or optics, infrared can be channeled into a narrow beam. That beam will travel farther than an unassisted source but will be fairly directional. This approach does have some drawbacks, but it will extend the reach of the light.

Another possibility, albeit an expensive one, is to use infrared laser diodes as your illumination. The cost of these lasers has decreased in recent years (they're used in compact disc players and writers, so the surplus market is reaping the benefit), but they do require dedicated power supplies. The result is a fairly complex lighting system, but it does have range—often in the several thousand-foot category.

One area where infrared light really shines is in underwater surveillance. Water has a nasty habit of absorbing light rather rapidly, so it gets dark quickly as you descend below the surface. In the same vein, visibility through water is not very good, so you will only be able to see well for short distances (fifteen to twenty feet, depending on the clarity of the water itself).

From our previous discussion of infrared LED light sources, it should be clear that IR is perfect for underwater observation. Infrared light penetrates water better than visible light. And since you can't see very far to start with, its limited range isn't as problematic. This is why you see most of the commercial underwater gear equipped with IR lighting.

SUPPLYING POWER TO THE EQUIPMENT

Let's talk a little about how you will install the actual equipment of your system. That equipment includes the camera(s), monitor(s), recorder(s), distribution system, and of great importance, power to run all these gadgets.

We have talked some about cameras and how they should be positioned. The monitors and recording equipment will be placed in an area designated for the purpose of monitoring the camera view(s). With most of this equipment, nothing special has to be done in terms of positioning to accommodate its operation.

Power, on the other hand, is going to be a major concern with any surveillance system. Both the camera and monitoring gear have to be powered. The monitoring equipment will probably have easy access to standard household 120 VAC outlets, but the cameras may be a different story.

Several factors make powering the cameras a little more complex than the other gear. For example, most of the cameras are going to be located high up, looking down. This provides the best view but does make getting "juice" to the camera, at best, inconvenient.

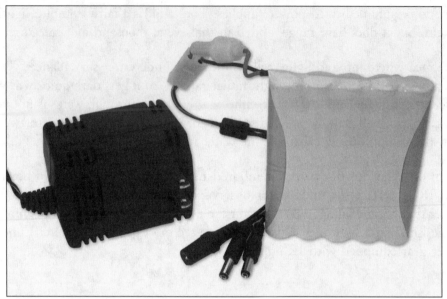

PHOTO COURTESY OF SUPERCIRCUITS, INC. USED BY PERMISSION.

FIGURE 11.4 A small, light-duty battery pack and charger.

Now, you can run the cameras off battery packs (see Figures 11.4 and 11.5). Most of the modern-day CCD or CMOS cameras are fairly frugal when it comes to power consumption, but eventually you are going to have to replace the battery pack. Thus, if this approach is used, be sure you can easily reach the battery.

A much more convenient method is to power the camera off standard house current. The cameras usually require five to twelve volts DC, and this is provided by the familiar "plug in the wall" power transformers. But, the wall-style power supplies do need access to the household power.

Here, you may need to contact an electrical contractor to provide this access. Since you will be dealing with power lines carrying dangerous electrical potentials that could result in fire if not properly installed, it might well show a degree of intelligence to let an expert do the job. These lines will probably be run through the attic region of your home, where they are not visible, and that in itself presents a hazard. Additionally, local regulations may require you to hire a contractor. Check with local authorities.

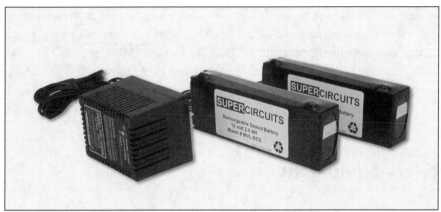

FIGURE 11.5 A battery charger and heavy duty batteries.

If there are no local restrictions and you feel comfortable doing the job yourself, then have at it. But, personally, I'm not inclined to recommend that anyone not trained in this area run 120 volts of anything through the attic. The bottom line is that the cameras will have to have power. Consider what is the best and safest method to provide that power.

Getting the camera's signal to the monitoring station is another issue. Here, you have two choices: hard-wired and wireless. I will cover wireless devices in more detail momentarily, but first, I'll discuss the hard-wired approach.

Hard-wired Equipment

In this scenario, the signal is carried to the monitor via a cable. The cameras are probably positioned high and the signal lines most likely run through the attic. The signal will be diminished by the cable itself, so cable length is a factor. If you want to install amplifiers along the cable run to boost the signal, that will work. But remember, the amplifiers will also require power. The average camera furnishes a strong enough signal to travel at least 100 feet down a cable. Unless you have some mighty remote camera locations, that strength is probably going to get the signal to the monitoring station without the need of an amplifier.

These lines are not going to carry high voltages and thus are not extraordinarily dangerous. The output voltage of all NTSC video cameras is one volt peak to peak, and that's unlikely to constitute a fire hazard. However, NTSC camera output signals are rated at seventy-five ohms, and cable to match should be used. This would be the RG-59-type television cable that is readily available from electronics outlets as well as construction supply centers.

Wireless Equipment

For most surveillance systems, even in businesses, hard-wired cable distribution is probably the best way to get the camera signal to the monitoring location. But there may be times and/or camera positions that simply don't lend themselves to cable feeds. In this situation, the wireless approach may be the only answer.

Wireless transmission of video signals is covered at length in Chapter 8. When a camera location is not conducive to hard wiring, wireless is a very practical method of getting the signal to the monitoring area. However, there are some challenges involved in using this approach (see Figure 11.6):

- **Powering the transmitter.** Like the cameras and/or amplifiers, access to electricity is needed. Again, this can be done via existing household current.

- **Expense.** In addition to the transmitter, you often need a receiver at the monitor end. This equipment has dropped in price over the years and is no longer the financial burden it once was, but it does increase the overall cost of the system. But sometimes you don't have a choice.

- **Interference to other radio frequency equipment.** The Federal Communications Commission takes a dim view of low-power wireless equipment causing interference to other gear, especially licensed stations. You have to be sure your TV transmitters don't cause such interference. If they do, you may have to turn off your gear. In the end, that could mean you bought a lot of expensive equipment for nothing.

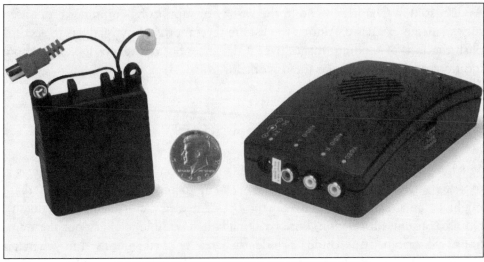

PHOTO COURTESY OF SUPERCIRCUITS, INC. USED BY PERMISSION.

FIGURE 11.6 A wireless transmission link.

- **Unauthorized reception of your video signals by other people.** Remember, you are sending a radio signal out over the air and anyone with the proper equipment can, and has the right to, intercept that signal. This could land you in a predicament involving making the surveillance information public. Just something to ponder. Additionally, you probably don't want other people receiving your surveillance signals. Particularly regarding your home, the surveillance is for the protection and safety of your family and property, and not really intended for "public" consumption.

REMOTE LOCATIONS

There are times, especially in business applications, where remote surveillance may be necessary, or it might just be a matter of convenience. In either case, you have to arrange for a method to get the signal from the camera(s) back to a location where it can be monitored.

As discussed in Chapters 8 and 9, the variety of supporting equipment for video surveillance is extensive. Among the list are several prospective gadgets to accomplish the task of sending images over a distance. So let's take a look at remote locations and how to make them work for you.

Purposes of Remote Observation

Much like any other form of video surveillance, the purpose of remote cameras is to watch areas that are otherwise inconvenient to monitor. For instance, there might be an isolated warehouse on the back lot of a factory. This area doesn't require constant observation, but occasionally it is sensible to check out the warehouse. Security guards could include the area in their rounds, but when an unscheduled inquiry becomes unavoidable, a guard has to be pulled from normal duty to investigate the problem. This is not only time consuming, but also disrupts normal security at the plant. The answer might well be to install several video cameras in and around the warehouse. In that scenario, a constant watch exists for whatever situations arise. Security people don't have to expend valuable time or concern themselves with a secondary section of the factory until there is a need. And, when a need does occur, the video equipment is there to immediately take up the slack.

This is, of course, only one example of how remotely positioned cameras can be of assistance to a business. In any circumstance where a part of the overall operation is located in an isolated area, or where it's germane to check the area only periodically, remote surveillance is a prudent option.

The same can be said for the home, but usually on a smaller scale. Attic recreation rooms, poolside areas, garage living quarters, or any other part of the home that is separated from the main structure can easily be monitored with video cameras. This approach not only allows for observation of activities in those areas but also provides an added safety factor (and peace of mind) in case a problem arises (see Figure 11.7).

All in all, surveying remote locations can be a judicious move. Once the initial investment is made and absorbed, the cost of operating such a full-time system is next to nothing, especially when compared to hiring a full-time security guard. This approach to business and home protection is well worth investigating.

PHOTO COURTESY OF SUPERCIRCUITS, INC. USED BY PERMISSION.

FIGURE 11.7 A remotely controlled camera.

Cost Investment

The initial investment depends on the scope of the operation. In general, many systems can be installed at a very reasonable—and maybe even surprisingly low—cost. Cameras have become almost ridiculously priced, and they are one of the major expenses. Black-and-white board cameras, as of this writing, are available for $30 in single units, and in bulk can be purchased for considerably less per unit. This helps make the overall investment far more attractive.

Signal Transmission

In addition to the cameras, a method of transmitting their images is required, as discussed earlier in this chapter. The hard-wired approach provides a very secure and reliable technique of moving the camera signal from the remote location to the monitoring station. However, while this method might at first glance seem to be the least expensive approach, in the end, it might not necessarily be the cheapest way to go. Hard-wire connections demand coaxial cable, which costs fifty to seventy-five cents a foot, and it might take several thousand feet of cable to do the job.

Another popular option is to use standard telephone lines. There are a variety of devices on the market for this purpose, most of which don't interrupt normal telephone operation. Actually, you can at least partially thank computers for this. The advent of the modulator-demodulator (modem) brought a realization of possible secondary applications of the telephone system. If you have a telephone line installed between the monitoring station and the remote area, you may well be able to benefit from this technology. Phone lines are a bit easier to tap into than dedicated hard wire, but in most cases tapping surveillance lines is not a top priority. However, if in your particular situation sensitive images are transferred over the system, this could be a definite consideration. Costs for this equipment have decreased dramatically in recent years.

The last method of transmitting the pictures is, of course, wireless. Of course, if security of the transmission is a concern, wireless links may be a problem, as mentioned earlier. Wireless links are a convenient, and relatively inexpensive way to move video pictures from a remote location.

Along with the expansion of microwave frequency applications came an important advantage for the wireless and video surveillance industries. This range of the frequency spectrum offers some benefits that are unique to it and it alone. For one thing, there is little of the noise and interference experienced at other frequencies. Additionally, signals will travel line of sight practically forever on very small amounts of power.

All of this makes microwave a natural for our purposes. And this has not gone unnoticed by the equipment manufacturers. Nine hundred megahertz, 1.2 gigahertz and 2.4 gigahertz wireless systems abound on the video surveillance market. Add high-sensitivity receivers and specialized antennas and you will have a highly efficient, thus effective, communications link.

Naturally, for remote locations these systems are perfect. As long as signal interception is not a problem, the easiest, and maybe in the long run cheapest, solution to especially distant remote sites is wireless. Initial costs have plummeted in just the last few years and transmission systems that carry several miles can be had for under $600. Considering the tens of thousands of dollars you would have shelled out for a comparable system just a decade or so ago, things have come a long way in our direction.

Complications

Cost, distance from the main structure, and possible tapping are the primary complications involved in the various methods used for remote location monitoring. But, as we have seen in this discussion, they are all either adjusting to market demand, as in cost, or easily overcome.

One complication that we have not touched on encompasses maintenance of the remote equipment. In this day and age of semiconductor gear, unusual maintenance or equipment failure is far less likely than it was in the days of vacuum tube devices. This should not be a particular obstacle with most remote operations. From time to time, though, there are going to be unexpected difficulties, such as power surges or lightening strikes that can throw you a curve. With the proper surge and lightning protection devices, normally this is not a routine dilemma. It is something to consider for any of the methods described for remote locations. However, an occasional requisite visit to the remote site for repairs should be more than justified by the expedience provided by isolated video capability.

CONCLUSION

Compared to what was available just a few years back, the equipment to monitor a site that is otherwise inconvenient to watch has experienced a genuine growth spurt. The opportunities are almost limitless.

12

Protecting and Maintaining Your Surveillance Equipment

INTRODUCTION

In this chapter, I will provide ways to protect and maintain your surveillance equipment. Some of this information has been covered elsewhere in the text, so let's regard this chapter as something of a review.

I hate repeating myself, but this is an important area, because reviewing methods of protection and maintenance could save you some problems and money. It would be bad for you to miss an opportunity to protect your home or business or not be able to identify an intruder because of an equipment failure caused by gear not properly protected or maintained.

PREVENTING DAMAGE—NATURAL OR OTHERWISE

When it comes to damage to equipment, many elements enter the picture. There is intentional damage from people, as well as unintentional injury from individuals and sources such as animals and the weather. Unquestionably, the best protection you can provide is **prevention.** Devote some forethought to placing your equipment and enclosures to prohibit easy gear access.

Probably the most vulnerable items in your system will be the video cameras. They tend to be more accessible than other equipment such as monitors, are often the target of vandalism, and can be susceptible to weather-related demons such as humidity, mold, mildew, dust, and extremes in temperature. This, of course, is especially true for cameras in exterior locations.

Fortunately, camera manufacturers are very aware of this problem and have taken appropriate protective measures with many of their camera designs. For example, a typical board camera (Figure 12.1) is frequently encased in a hard shell (Figure 12.2) or a vacuum-molded cover (Figure 12.3).

These are intended for indoor use only, but the covers prevent the delicate electronics from being damaged by handling, probing, or static electricity. The hard cases often come with a small mounting bracket that facilitates placement of the camera.

PHOTO COURTESY OF SUPERCIRCUITS, INC. USED BY PERMISSION.

FIGURE 12.1 A typical board camera.

FIGURE 12.2 A hard shell housing.

FIGURE 12.3 A vacuum-molded cover.

Many of the housings provided by equipment companies are overt in nature, but they do provide excellent camera protection. This might include the metal square cases, like the one shown in Figure 12.4, domes, such as the one in Figure 12.5, and surface mounts for ceiling installation, as in Figure 12.6. All will help keep the camera safe, as will plastic and metal interior housings (Figure 12.7 and Figure 12.8).

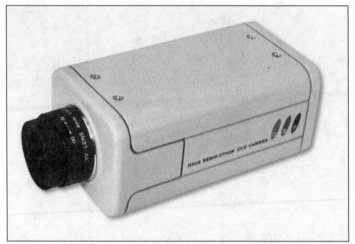

PHOTO COURTESY OF SUPERCIRCUITS, INC. USED BY PERMISSION.

FIGURE 12.4   Metal square cases.

PHOTO COURTESY OF SUPERCIRCUITS, INC. USED BY PERMISSION.

FIGURE 12.5   Domes.

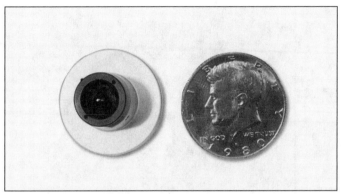

PHOTO COURTESY OF SUPERCIRCUITS, INC. USED BY PERMISSION.

FIGURE 12.6   Surface mounts.

PHOTO COURTESY OF SUPERCIRCUITS, INC. USED BY PERMISSION.

FIGURE 12.7   Plastic interior housing.

Additionally, indoor cameras can be housed in a variety of covert locations such as inside lamps, behind pictures, and in fake books on a shelf. Also, overt housings are available that help protect the boards, like the ones shown in Figure 12.9 and Figure 12.10.

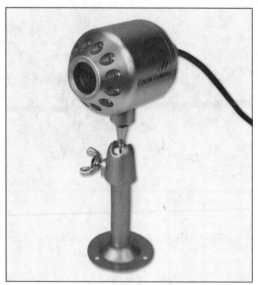

PHOTO COURTESY OF SUPERCIRCUITS, INC. USED BY PERMISSION.

FIGURE 12.8 These are obviously security cameras, but the cases protect the delicate electronics..

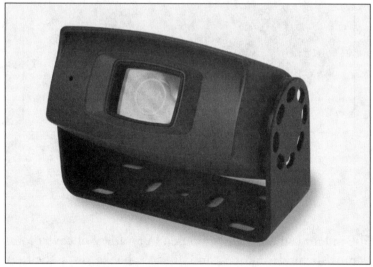

PHOTO COURTESY OF SUPERCIRCUITS, INC. USED BY PERMISSION.

FIGURE 12.9 Overt housing.

PHOTO COURTESY OF SUPERCIRCUITS, INC. USED BY PERMISSION.

FIGURE 12.10 Housings like these protect your board camera.

For outside locations, a number of safeguarding arrangements are available. In addition to the standard exterior camera housings that we covered in detail in Chapter 10, many manufacturers provide board cameras with outdoor-ready cases. Some examples are the combination camera/infrared lighting assembly in Figure 12.11 and the stand-alone exterior camera seen in Figure 12.12.

The small cylindrical lipstick cameras are a big favorite, as they are quite weather-proof and come in either black-and-white or color versions (Figure 12.13). Another favorite for semioutdoor employment is the electronic version of the familiar door peephole (Figure 12.14).

The lipstick cameras are hard to hurt, either intentionally or from weather-related incidents, and the door camera is tough to damage and easier than a lens to clean if someone decides to paint the thing. I will touch on the latter subject in detail later.

As has been previously mentioned, antihumidity, antimoisture, antimildew (in short, anti-anything wet that could do harm or cause malfunction) cases are available from several of the manufacturers listed in the source list. These housings

FIGURE 12.11 Combination camera/infrared lighting assembly.

FIGURE 12.12 Stand-alone exterior camera.

PHOTO COURTESY OF SUPERCIRCUITS, INC. USED BY PERMISSION.

FIGURE 12.13   The popular lipstick camera.

PHOTO COURTESY OF SUPERCIRCUITS, INC. USED BY PERMISSION.

FIGURE 12.14   The peephole camera looks just like the peepholes you see in hotel room doors.

normally employ self-contained heating units that keep the interior atmosphere warm and dry. They are also great for protection against extremely cold weather.

Regarding inadvertent damage done by small children and animals, the key here is placement. Actually, you are likely to have the cameras located high in a room anyway, but if the cameras are within reach of curious fingers or paws, it might be time to rethink your camera placement. Higher locations give a far better view of a room and definitely keep small hands and sharp teeth off the cameras. Depending on how high we are talking, the placement may also discourage spray paint, which is something to consider when it comes time to plan the location of the cameras.

One last note regarding intentional vandalism: disguised cameras are a great way to minimize damage. Whenever you have a camera masquerading as something else, it is hard for potential wrongdoers to know what to damage. This includes painting the lens or taking the proverbial blunt instrument to the camera. Disguising your cameras can go a long way toward protecting them. See Figure 12.15, Figure 12.16, and Figure 12.17 for some examples of disguised cameras.

PHOTO COURTESY OF SUPERCIRCUITS, INC. USED BY PERMISSION.

FIGURE 12.15 Exit sign.

PHOTO COURTESY OF SUPERCIRCUITS, INC. USED BY PERMISSION.

FIGURE 12.16   A camera masquerading as an electronical outlet plate.

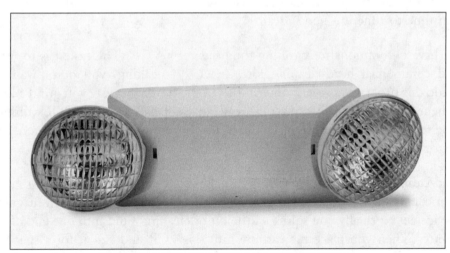

PHOTO COURTESY OF SUPERCIRCUITS, INC. USED BY PERMISSION.

FIGURE 12.17   If they don't know it's a camera,
they aren't likely to try to destroy it.

ROUTINE MAINTENANCE

What should come as great news is that most surveillance equipment requires very little in the way of routine maintenance. In this age of semiconductor electronics and printed circuit board architecture, the gear is extremely stable, reliable, and almost impervious to the problems that plagued the older vacuum-tube equipment.

Keeping the gear clean is important. A technician once observed that a camera that is dirty on the outside is going to be dirty on the inside as well. Dirt and dust can be problematic to all electronic equipment, and this stuff is no different. Manufacturers recommend that you periodically clean all the gear, especially exterior cameras, with a damp cloth, then dry. How often is periodically, you ask? Well, at the risk of sounding simplistic, clean them when they get dirty. It is a matter of checking the equipment from time to time and taking care of it when it becomes soiled or dusty.

One piece of equipment that does benefit from routine maintenance is videocassette recorders. VCRs have magnetic recording heads that will, with time and use, become dirty. Running a head-cleaning maintenance procedure on them from time to time is a good idea.

You have two options for cleaning the heads of a VCR. The easiest is to use a good commercial head-cleaning kit. Generally speaking, you don't want to overdo this program, as every time you clean the heads this way you grind a little of them away. Eventually, you will wear out the heads and the VCR will have to be replaced. You could replace the heads, but you would spend more doing that than buying a new VCR.

Head-cleaning kits are available from just about any source that sells VCRs. They often come as a videocassette look-alike, with a cloth strip rather than magnetic tape inside. Normally you squirt a little bit of cleaning fluid into one end of the cassette. When you place the cassette into the VCR and run it for a specified time, the cleaning fluid-soaked strip scrubs the dirt off the heads.

Another option is to open the VCR case and use acetone and a soft lint-free cloth to clean the heads and denatured alcohol to clean the rollers and plastic pieces.

You do have to open the unit to do this but it does greatly increase the life of the video heads. You can also, of course, pay $30 to have a certified professional clean it for you once in a while if you don't want to void the warranty by opening the case.

Beyond that, there is really not much you can do in the way of maintenance. All of this equipment is definitely not consumer serviceable, so in the event something acts up or breaks, take it to a manufacturer-recommended repair technician. Most likely, though, you will find that it is faster, cheaper, and more convenient to simply replace broken gear.

CLEANING UP AFTER VANDALS

I have often been asked what procedure should be used to remove spray paint from a camera lens. Unfortunately, lenses tend to stick out and are frequently the target of anyone trying to disable your surveillance system. Even more sadly, there is little you can do to fully restore a painted lens. (See Figure 12.18 and Figure 12.19.)

My advice is to remove the lens from the camera, avoid getting any wet paint on your hands, and throw the little gem away. If the paint has had time to dry, you won't have to worry about getting paint on anything, but you surely have a lens that will be extremely hard to restore to a useful condition. The good news is that the cost of remarkably high-quality lenses has really dropped in recent years.

I know this probably sounds wasteful, but lenses today are often made with plastic components, and the solvents required to remove paint are notorious for melting, or at least softening, plastic. If the optical elements are plastic, they are likely to be fogged or damaged by said solvents. If the elements are glass, they still have an optical coating that most solvents will remove. Either way, the result is degradation of the picture quality.

Additionally, many of the mounting assemblies that hold the optical elements are made of plastic. The same melting or softening problem applies to these assemblies. These mounts are usually not watertight, and that means that when you

PHOTO COURTESY OF SUPERCIRCUITS, INC. USED BY PERMISSION.

FIGURE 12.18 Zoom lens.

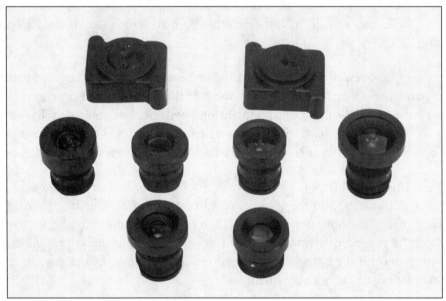

PHOTO COURTESY OF SUPERCIRCUITS, INC. USED BY PERMISSION.

FIGURE 12.19 You may need to replace a lens from time to time.

attempt to clean the lens, some of the solvent is likely to seep inside the lens assembly. This, with time, can cause the elements to change position and/or fog and both conditions render the lens useless.

I repeat my original advice: throw the thing away! However, if you just can't help yourself and are determined to try and rescue the lens from a paint attack, here are some tips.

First, it will be necessary to determine what type of paint was used. Was it enamel, latex, or lacquer? This is important, as the solvent you need depends on the paint type. Using the wrong solvent can make matters even worse. Unfortunately, once the paint has dried, they all look pretty much alike. The best way to figure out what you've got is by trying different solvents on a small area to see what happens.

First, a safety note. **Do not** use carbon tetrachloride for removing any kind of paint. While this stuff is nonflammable and a great paint solvent, it is extremely toxic. Just spilling the chemical on your skin can cause toxic hepatitis. Stay away from carbon tet, which is hard to even find anymore.

For enamels, denatured alcohol, paint thinner, and mineral spirits are a good place to start. Apply the solvent and clean the lens as carefully and gently as possible. Remember, the surface of the exposed optical element has been polished to a high degree, and any changes or distortions in that polished surface will affect the performance of the lens, sometimes quite dramatically.

If the enamel has dried, it may be stubborn. The next solvents to try would be acetone and xylene. Be careful with both of these. They are notorious for doing a fine job of melting plastics and fogging clear plastics. They will also take the paint off of anything they come in contact with, which could lead to other difficulties.

Another safety note. All the solvents I have mentioned are both toxic and flammable. They should only be used in a well-ventilated area, should not come in contact with bare skin, and should be kept completely away from any open flame. I particularly stress this last item. You don't want to end up looking like Wiley Coyote after one of his failed attempts to blow up the Roadrunner.

For latex paint, if it has not dried, water will often do the job. This, of course, is the best of both worlds. Water is not flammable or toxic, and it won't soften or fog plastic parts. It will, however, rust iron-based metals, so be sure to remove any residual water after cleaning the lens.

If the latex has dried, the solvents mentioned above may take it off. Again, start with the milder ones and be gentle.

Lacquer is a tough one. Use only lacquer remover. Anything else will get you into big trouble. Try to carefully remove the paint, but dried lacquer is hard to soften up and even harder to easily remove. Lacquer remover is both toxic and flammable, so use appropriate care in handling it.

DAMAGED COMPONENT REPLACEMENT

As I said earlier, just about all the video surveillance equipment you will encounter is not consumer serviceable. Even if you are very adept at electronics, the nature of construction, with its tiny surface-mount parts, and the complexity of the circuits make the gear next to impossible to repair yourself.

Also bear in mind that the cathode-ray tube (picture tube-type) monitors have **very high voltage** supplies inside, and, trust me on this as I speak from personal experience, you do not want to tangle with those potentials. It probably won't kill you (I'm still breathing), but you will be rather sorry you even opened the monitor's case.

When trouble looms its ugly head, it is often prudent to merely replace the gear. It will most likely be cheaper to do that than to pay the exorbitant service fees charged by many repair facilities. Much of this equipment has come so far down in price that it is almost disposable.

I have talked about replacing damaged or vandalized lenses, and the same can be true of such things as cables, such as the ones that connect the cameras to the monitoring gear, mounts, and housings that become damaged. The cable is normally coaxial in nature and over time can become frayed, crimped, or broken. Usually degraded picture quality lets you know when the cable needs replacement.

Camera mounts or cases, especially when exposed to the elements, are another area to watch. The normal contraction and expansion brought on by changes in temperature can cause housings to eventually crack. It is a good idea to watch for signs of this, as a cracked case can let in unwanted moisture. Plastic mounts are susceptible to the same pitfall, and a broken mounting bracket can cost you a perfectly good camera.

When something beyond the easily replaceable cables, mounts, and housings goes bad, complete replacement of the equipment is the better part of valor. The source list at the back of this book provides a variety of companies for this purpose, and competition is tough these days, so the prices can be quite good.

SCHEDULED EQUIPMENT REPLACEMENT

The good news here is that modern video surveillance gear is so reliable that you may never need to replace a lot of it. I can't really think of any bad news. Unlike the days of vidicon tube cameras with their 5,000-hour maximum life expectancy, charge-coupled device (CCD) imagers normally last just about forever (or, at least our lifetimes). Electronic-based problems may develop over the years, but the equipment cost is so reasonable that it is wise to just replace the gear.

If anything might fall into this category, it would be videocassette recorders. If you put your VCRs on a regular head-cleaning routine (and **not** cleaning the heads is not really a good way to avoid the inevitable), eventually you will wear out the heads and the VCR will have to be replaced. You could replace the heads but you would spend more doing that than buying a new VCR. One other area to watch is the videotape itself. Depending on the quality of the tape (and I don't recommend using anything but the best quality tape), it will wear out with time and use. Again, when the picture quality begins to consistently suffer, it is probably time to buy new videotape.

CONCLUSION

Hopefully this review has been useful to you. As I have said repeatedly in this chapter, the low cost of so much of this surveillance gear has made buying and using it a real treat. Not only does it enable you to install a system of quality gear, it provides for relatively painless expansion, at least as far as your pocketbook is concerned. Additionally, it makes replacement of equipment less painful. Take the precautionary steps that we have discussed with gear that requires attention, but most of the other stuff takes care of itself.

13

Countersurveillance

Where there is surveillance, there is going to be countersurveillance, especially when the surveillance is of the clandestine variety. If someone is listening to or watching you without your knowledge, wouldn't you want to put an end to it, or at least be aware of the surveillance? I think most people would answer yes to that question.

There are ways to detect and locate surveillance devices. In fact, for those with the knowledge and equipment to "sweep" rooms, offices, buildings, and other property, a very profitable business is at hand. There is a procedure to follow, which will be discussed in this chapter, and the results, even with basic equipment, can be quite effective.

DETECTING HARD-WIRED DEVICES

While hard-wired (connected by cable) clandestine surveillance equipment is becoming rarer, there are still scenarios where it is the best way to go. In situations where the bug is installed on a permanent basis and the installer has adequate access to the building to hide the cable, the hard-wire method is often preferred. Also, this approach allows power, if any is needed, to be taken from the normal building AC lines.

Another advantage to hard wiring, especially with microphones that don't require power, is the difficulty in finding the installed gear. In an arrangement where a cable carries the signal, there is no radiated signal to be encountered by a bug detector. Some of the very sensitive, and expensive, detectors can actually pick up electricity moving within a circuit, such as in an amplifier, but that, at best, can be a questionable matter.

If the circumstances are right regarding necessary structural access, a hard-wired system will be difficult to find. For example, if the occupants of a building anticipated the need for listening microphones in certain parts of the building, they might well have had the necessary wiring put in place when the structure was constructed. In that case, it is going to be very hard to uncover those bugs.

However, in many scenarios the listening devices are "second thought," and the wiring is run down wall seams and along the edge of carpeting on the floor. Normally, people who know what they are doing can conceal the cable well enough to keep it from being spotted. On close scrutiny, however, the wiring will be discovered, revealing the presence of the device.

Often, this is the only way you will spot the device. Naturally, once the cable has been found, it can be traced to the microphone, video camera, or whatever else might be connected to it. In short, if time and access to the structure exists, a hard-wired bug is going to be more detection proof than a wireless system.

DETECTING WIRELESS DEVICES

A more frequently encountered method of bugging, either audio or video, uses a radio frequency (RF) transmitter to send the sound and image out over the air and subsequently to a receiver that performs the monitoring function. This arrangement does require additional equipment, but it's far easier to install. The time factor can be important when placing a clandestine bug. It often has to be done in a matter of seconds or minutes at the most, and the simplicity of merely having to hide the transmitter in an unobtrusive location makes wireless systems very attractive. I hate to nag, but remember, covert surveillance, especially of someone else's property, is normally illegal.

However, the fact that an RF source is present also makes these varmints considerably easier to detect. So much so that simple wideband receivers, if placed close enough to the bug, will usually pick up their signal. The compromise has to be accepted that wireless is easier to install but also easier to find.

One of the projects in Chapter 15 is a simple but effective bug detector. This unit only detects RF bugs, but those are by far the most commonly seen devices. Incidentally, there are a few systems that employ modulated laser light to transfer information, but at this point these are pretty much experimental in nature.

When working with an RF radiation detector, it is best to start at the top of the room and work down. Be very meticulous as you move around the room and down the walls, as some of the bugs are so low in power it is easy to miss them if the sweeps are too wide or rapid. Once the walls are declared clean, begin moving toward the center of the room. Do this at the roof, then at the floor level.

Check everything! As shown in Chapter 9, "Special Purpose Cameras," both audio and video bugs can be hidden in lamps, books, pen sets, videotape containers, pictures on the wall, and you name it. Desks are especially vulnerable to bugging as they contain so many compartments and other inviting places to place the little gems. If you can reach the center of the room at both levels and still not detect a transmitter, the room is probably clean. (It really wouldn't hurt to go back over it all once more just to be sure.)

Also, several sweeps over several days is a wise approach. Some bugs have become very sophisticated and often can be turned on and off remotely. They also might turn themselves on and off depending on activity in the room. This is known as *voice activated* or *VOX* and is employed to conserve the device's battery.

Wireless is a going concern these days and there is little evidence to suggest that will change in the near future. If you have even modest equipment and the skills described here, you should be able to find any RF bug placed in your home or business.

In the way of places to get equipment, the source list at the back of the book is a great place to start. Rather than mention specific models, let me talk about what is available, how it works, and what works the best.

Naturally, the more you are willing to spend on detector equipment, the better it will work. At the high end, a spectrum analyzer is a multifrequency receiver, covering from just off the AM broadcast band, or 1600 kilohertz, all the way up into the microwave range, 10 gigahertz and sometimes higher, that is connected to a dedicated oscilloscope. An oscilloscope is a device that displays, usually through a cathode-ray tube, various data in a very precise manner. This is the gadget you see in sci-fi movies with a small round screen that has wave patterns running across the screen. The receiver/oscilloscope combination can not only accurately detect the presence of a transmitted signal, it can also tell you what frequency it is operating on. The spectrum analyzer is a very useful device, but they start at around $4,000 for just a basic unit. The good ones can come at better than $10,000.

On the other end of the budget, a homemade device like the one in the project section will do an adequate job of finding RF bugs, but its sensitivity level does necessitate your getting close to the radiation source, that is, the bug. If you are able to spend a few hundred dollars, you can get a detector that doesn't have to get quite so close to catch the transmitter.

That in itself can be good or bad, and it is often wise to have both a high-sensitivity and low-sensitivity detector in your bag of tricks. The reason is simple. The high-sensitivity device will be able to detect the presence of the bug, but because it does so at a great distance, it is hard to pinpoint the transmitter. That is where the less sensitive detector comes into play. Once you know there is a radiator in the room, the low-sensitivity unit will allow you to move right in on it.

Many of these detectors feature both audible and silent modes, which can be very useful. Often, it is not in your best interest to let the person who has planted the bug know you have found the device. You may want to pass along some bad information, if you know what I mean. In that scenario, you would not want to tip the perpetrator off with any type of audio detection signal from the detector. More about that in a moment.

Another handy-dandy device, know as a frequency counter, proves useful when hunting hidden transmitters. These units not only detect the signal, they also tell you what frequency the transmitter is operating on. That information can sometimes be very useful in identifying the owner of the bug. For example, some frequencies are going to be used only by official bugs, such as those planted by law enforcement, while others may be employed by more conventional devices, such as the variety of small, portable transmitters available on the open market.

In the way of a little trivia, the name frequency counter is a misnomer. In reality, these devices are a specialized event counter. That is, they have what is known as a timing window, or gate, circuit that allows the counter to operate just for a very specific period of time. The number of times an event occurs during that window becomes the counter measurement. For example, let's consider one cycle of a radio wave as an event. If the window is one second, and the counter detects one million events during that second, then the frequency of the source will be one million hertz, or one megahertz.

Anyway, that should give some idea of how the frequency counter functions. Trust me on this, if you plan to do much in the way of bug detecting, a frequency counter is a prudent investment. A good one can be purchased for under a hundred dollars these days. They may be more expensive if you require a higher frequency limit—that is, how high up the frequency ladder they can count.

SHOULD YOU REMOVE THE BUG?

Suppose you find an audio bug in your office. What do you do next? Tearing it out isn't always the best route to take. If someone has taken the time, expense, and effort to bug your office, there must be something going on in your office worth listening to. Most of the really important decisions about your business are probably made within that office, and a competitor would probably love to hear those decisions. If your office has been bugged, someone really wants to know such things as what your new product line will look like, how your research is going, or what your future business plans are.

If you are unaware of the bug, that person might just get all that information. But, since you now know it is there, a little game played by the big boys (professional intelligence people) might well work for you. *Disinformation* is the art of feeding the scoundrel that placed the bug **bad** information. Someone may well pay an even higher price than what it cost to get the bug installed by making decisions based on the bad information you are providing. Kind of makes you feel warm all over, doesn't it?

The moral is that it doesn't always pay to smash that covert device upon detection. Sometimes it can be used to your advantage instead of for the person who installed it. It is just a matter of choice. If you do decide to destroy the device, a hammer is the best way I know. Beat that sucker into tiny little pieces and that should preclude any additional surveillance by that particular device. If in doubt, kill the thing, bury it, then dig it up and kill it again!

CONCLUSION

Do not forget to inform the police if you find and elect to destroy a bug. This type of intrusion into your personal or professional life is against the law. If you don't plan to put the bug to use for your own devices, then maybe you can put the person responsible for placing it in jail.

14

The Future of Surveillance

Before jumping into the Project section of this book, which gives you an opportunity to experiment with some of this great gear, let me briefly address what the future of surveillance gear might hold. A lot of this is speculation, but I don't think it gets too far off the beam.

Based on what I've seen in just the last twenty years, and the ingenuity of human technology, I foresee an even more amazing and effective line of devices. And the way that computer and digital technology is exploding, any discussion of digital recording and video enhancement technologies would be out of date before it gets into print! In short, I don't see any end to the parade of astounding equipment.

THE INCREDIBLE SHRINKING GADGET

I think size will continue to dominate the research and development world. Audio transmitters have certainly gone the miniaturization route, and I predict video cameras will take the same path. The evolution of cameras too small to be seen with the naked eye would not surprise me, and the Dick Tracy wristwatch TV transceiver is practically a reality.

Smaller everything will be the order of the day. Of course, there does come a point when too small becomes impossible to use, but if past experience is any gauge, manufacturers will find a way to overcome that obstacle. Look what they did with those tiny pieces of silicon we all call integrated circuits. Covert cameras are so small that they are currently being installed in fully functional wall clocks, thermostats, smoke detectors, dummy sprinkler heads, books, and pinhole cameras that can get 360 degree view through a one-eighth-inch hole.

Fiber optics might well see additional use in the surveillance field. While still pricey as of this writing, it is probable that fiber optics will take the same path much of technology is taking. That is, it will become more abundant, drop in price, and improve in quality. Quality will largely determine how useful fiber optics become to video surveillance. The concept is very useful with sound, though, and certainly has the necessary bandwidth, or ability to carry a high frequency range, to accommodate video signals. Fiber optics are used quite extensively now in high-end and high electrically noisy applications. Watch for fiber optics to play an even more important role in transferring surveillance signals in the future.

QUALITY IMPROVEMENTS

I also see the quality of cameras improving to an almost unbelievable point. Resolution for even color cameras is approaching broadcast quality (525 lines or better) and that is for the lower-end devices. Postage stamp cams with 600 lines of horizontal resolution may be only a few years away.

Also, liquid crystal displays (LCDs) will improve many times over in the next few years. They are darn good now, but they will get better yet. All of this will eventually lead to highly portable, very high-quality surveillance systems that can be carried to wherever they're needed just like cell phones or pocket radios.

One last prediction involves wireless devices. I can clearly see transmitters having better frequency control, longer range, and cheaper prices. Right now, if you catch a good sale, you can pick up a 900 megahertz camera and monitor system for under a hundred dollars. As we have seen in the book, transmitter sizes have already dropped to near ridiculous dimensions and the range is getting better all the time. In the not-too-distant future, I suspect you will see a TV camera, transmitter battery, and antenna all in a package the size of a matchbox. And it will have enough range to reach the nearest satellite.

CONCLUSION

Well, so much for some of my prophecies. I'm sure you have a few of your own. In all seriousness though, this field is in for tremendous growth and greater application in both the home and business arenas. At the risk of alarming readers with a mental image of "big brother," the cost, reliability, and size of video equipment will allow for cameras to be placed in all sorts of places they have never been seen before. This, in all fairness, may not be as bad as it might sound at first.

Have fun!

PART FIVE

Ready to Try?

Here comes the fun stuff! The first chapter in this section provides some relatively easy projects—switches, antennas, modulators, bug detectors, and more—that take just a weekend (or less) to build, and give you some useful surveillance technology.

The final chapter provides some additional information about two technologies that are really experiencing some exciting growth in the surveillance field—video recorders and Liquid Crystal Display monitors.

15

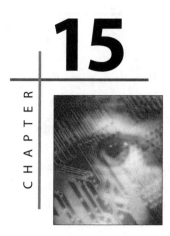

Projects

This chapter presents six projects that may well come in handy for your home or business surveillance setup. All are designed to be easy to understand and construct, so they shouldn't give you any problem if you decide to tackle one or more of them.

All are devices that would enhance various aspects of a surveillance system. Each can be built over a weekend, in some cases in a single night, and the end result will perform a needed surveillance function. Whether it be utilizing a standard TV as a monitor or getting better range out of your wireless system, these projects offer you something useful.

AUTOMATIC TELEPHONE LINE TAPE RECORDER SWITCH

Let's start out with a handy device that will allow you to record your telephone conversations. This one is a kit, and could definitely be of enormous value in certain situations.

This is a kit from The Electronic Rainbow that allows you to tap your own telephone. With the device on your phone line, you are able to make an audio tape recording of both sides of the conversation for every call, incoming or outgoing, that takes place on that line. Virtually any recorder with separate mic input and tape activation on/off jacks can be employed.

A quick look at the schematic diagram in Figure 15.1 reveals a very simple circuit. The component count is minimal at seven, and the theory is just as simple. First, a brief explanation of how your telephone works. When your phone is on-hook, the voltage across the two input lines, called Tip and Ring, is about forty-eight volts direct current (DC). However, when you answer the phone or pick the receiver up to make a call, creating an off-hook condition, that voltage drops to somewhere in the neighborhood of five volts DC.

What is needed to activate a tape recorder is a switch that detects that voltage drop. In this project, that detection is accomplished by resistors R1 and R2 and transistor Q1. The resistors are configured as a voltage divider that keeps Q1 conducting, not allowing enough voltage to reach the base of transistor Q2. That condition keeps Q2 nonconducting.

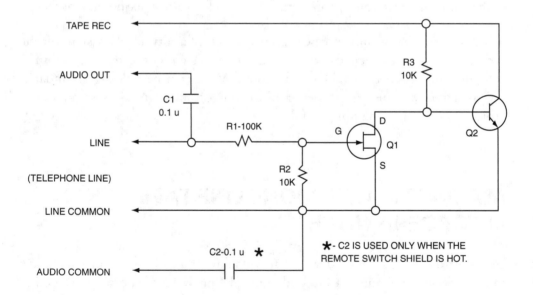

Q1: 2N7008 N-CHANNEL MOSFET TRANSISTOR
Q2: MPSA 14 NPN SIGNAL TRANSISTOR

FIGURE 15.1 Telephone auto-record switch schematic.

When the telephone line experiences the voltage drop from the off-hook state, it causes Q1 to stop conducting. In turn, Q2 starts conducting and that activates the tape recorder. The phone conversation is fed to the recorder's mic input jack directly off the line, which means it is recorded onto the tape.

Capacitors C1 and C2 act as DC blocks to capacitively couple the audio to the recorder. That is all that is needed to make an audio copy of what could be heard during the telephone exchange. As might be expected, that recording could be very useful in a number of ways.

As for construction, simply follow the instructions with the kit. A printed circuit board (PCB) is included, and it is just a matter of soldering the parts to the proper place on the board. Once completed, the PCB can be enclosed in an appropriate case. In case you want to wing it, a PCB pattern is included in Figure 15.2, and Figure 15.3 shows the stuffed PCB.

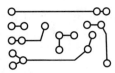

FIGURE 15.2   Telephone record switch PCB pattern.

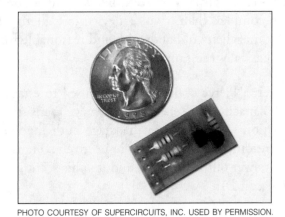

PHOTO COURTESY OF SUPERCIRCUITS, INC. USED BY PERMISSION.

FIGURE 15.3   The stuffed PCB for the automatic telephone recorder switch.

A length of standard telephone cable is used to connect the switch to the phone line, and one of the two mic input jack configurations is used to connect to the recorder's input. The tape recorder cable hooks to the recorder's on/off jack. Depending on the tape recorder you use (see Figure 15.4), these are either 3/32 or 1/8-inch phone plugs.

Using the device is as simple as connecting it to your recorder and telephone system. Put a tape in the recorder, and set it for Record. When the phone goes off-hook, the recorder starts, and the conversation is archived for posterity.

WIRELESS TELEVISION TRANSMITTER

This device (see Figure 15.5) can come in handy for all sorts of purposes. It is a completely self-contained video transmitter with a range of between 100 and 300 feet depending on the surroundings. Additionally, it is tunable to the lower VHF portion of the television broadcast band, channels two through six.

Among the ways you might use this are as a video babysitter, in a remote surveillance location, or as a covert monitoring camera. In one highly portable package you have a black-and-white or color camera, VHF transmitter, and battery.

While a TV transmitter might at first seem like a complex project, in reality it is not much more complex than a voice device (see Figure 15.6). The only requirements are six megahertz of bandwidth and reasonable frequency stability. The principles are basically the same.

With our transmitter, a Pierce oscillator (Q3) is used to establish the operating frequency. Variable capacitor C7 sets that frequency. Transistors Q1 and Q2 are used as the modulation section, and in this stage both the video gain and DC bias are set by potentiometers R1 and R2, respectively. Transistors Q4 and Q5 are configured as a power output section, and it's this section that is inductively modulated through coil L1.

The result is an efficient, yet simple, transmitter that will provide video images on normal commercial TV channels two through six. This aspect gives the

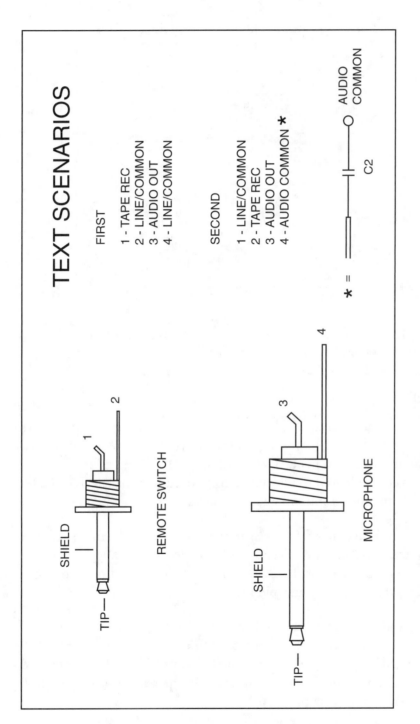

FIGURE 15.4 Plug arrangements for the telephone record switch.

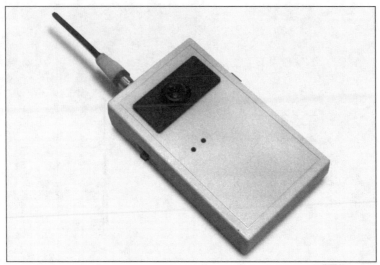

PHOTO COURTESY OF SUPERCIRCUITS, INC. USED BY PERMISSION.

FIGURE 15.5   The completed pocket TV station. The case holds the camera, battery, and transmitter, and has a range of between 100 and 300 feet, depending on surroundings.

transmitter a great deal of versatility, but also makes unauthorized viewing more likely. However, the relatively short range of the device does help keep the signal from being monitored by too many TVs. With cable TV being as popular as it is these days, fewer people have actual antennas hooked to their televisions, and an antenna is a must for reception of this broadcast.

A PCB makes construction of the transmitter easier. The board pattern is supplied in Figure 15.7, and it is wise to use a double-sided arrangement. That is, have copper on both sides of the PCB. This provides a good ground plane, which leads to better frequency and picture stability.

The easiest way I found to make the PCB was to etch the pattern side then drill holes in the appropriate places (see Figure 15.8). On the reverse side, use a small hobby tool to grind off the copper around the holes, unless a hole is connected to ground. Grind enough of the copper to prevent a possible short to ground, but leave as much copper as is practical.

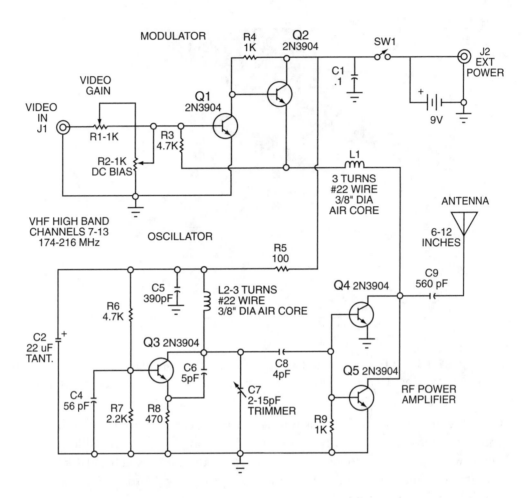

FIGURE 15.6 Television transmitter schematic.

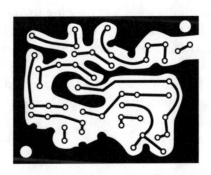

FIGURE 15.7
TV transmitter
PCB pattern.

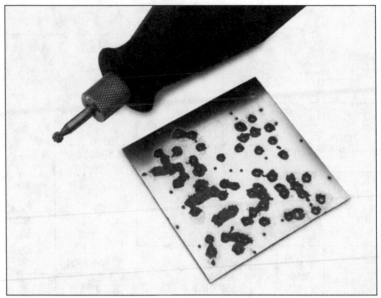

PHOTO COURTESY OF SUPERCIRCUITS, INC. USED BY PERMISSION.

FIGURE 15.8 The completed TV transmitter PC board. Notice the hobby grinder that was used to grind copper from the component side of the board.

Once the printed circuit board is finished, it is merely a matter of soldering the components to their respective positions. As always, solder both sides of the board whereever a component lead is tied to ground, and watch out for solder bridges and cold joints. Properly done, this process will produce a small but sturdy bundle (see Figure 15.9).

Packaging is really a matter of personal preference. I used a small (4.5 x 2.5 x 1-inch) project case that has a door on the back for changing the nine-volt battery, but any suitable case will do. As for the antenna, I found a six-inch length of stiff wire worked very well. Be sure to cap the exposed end for safety purposes.

And that is about all there is to it. With everything in place and a set of rabbit ears hooked to the TV antenna input, tune your television to one of the channels two through six. Adjust capacitor C7 until the TV receives the signal, then adjust potentiometers R1 and R2 for the best picture. A very simple set-up procedure.

PHOTO COURTESY OF SUPERCIRCUITS, INC. USED BY PERMISSION.

FIGURE 15.9 An internal view of the pocket TV station.
A nice compact arrangement.

Once you plug the camera's video output into the transmitter's video input and the transmitter/camera is working, you have a highly portable surveillance device. You may also be the only person on your block with your very own television station.

SIMPLE VIDEO SWITCH

When I say simple, I mean simple—and effective! In reality, switching video is no different than switching audio (although, you will hear some people say differently). In that vein I have designed a device that will allow you to switch between three different video/audio inputs. This is very useful for quickly changing from one surveillance camera to another when only one monitor is available. As you become more familiar with its potential, you will surely come up with other applications for this video/audio switch.

In all seriousness, this switch will come in handy, if not now, then down the line. All that is needed is a commonly available 2P3T rotary switch, eight RCA jacks, a knob, and a metal case. The metal case will help prevent extraneous noise and RF interference from getting to the switch.

As you can see in Figure 15.10, the switch schematic, this project is a simple matter of wiring the center conductors of the input jacks to the rotary switch's throw positions, and wiring the center conductors of the output jacks to the rotary switch's pole positions. All the jack shield connectors are tied together as the ground rail.

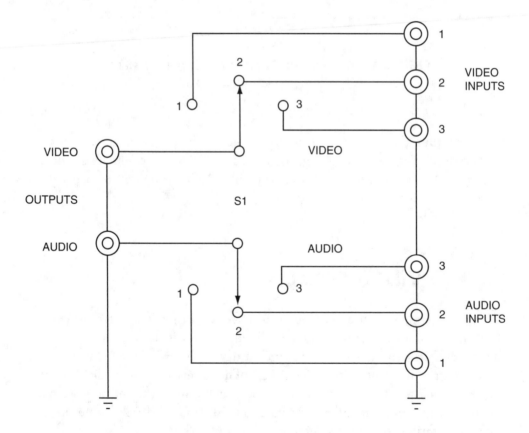

FIGURE 15.10   Video switch schematic.

Voilá, you have a three-position video/audio switch. When the jacks and switch are installed in a small aluminum or steel case, the requisite shielding will keep the signals moving through the switch without outside interference. I also advise that you label the case. That helps you keep the connections straight.

ULTRAHIGH FREQUENCY (UHF) BEAM ANTENNA

This is a device that will be useful if you are using the seventy-centimeter amateur radio band or some of the UHF shared frequencies. The design for this project is cut to the exact size for the 438 to 444 megahertz HAM channel (ATV), but it can be sized for other frequencies as well.

Here is some background. This type of antenna is known as a *Yagi* and is named after its Japanese inventor. Yagis consist of three types of elements: reflectors, driven elements, and directors. Looking at Figure 15.11, you can see that the reflector (always the longest element) is on one end of the center support or boom. The next element is the driven element, and this is where the actual signal is applied.

Following the driven element there will be the director, or in many cases, the directors (plural). In our design, three directors are used, but Yagis can have ten, fifteen, twenty, or more director elements. The more directors in the system, the more directional the antenna is going to be, and the stronger the received signal.

Like all antennas, the length of the elements must correspond to the frequency it will be operated on. In our case, the elements are cut for 439.25 megahertz, or the center video frequency for the ATV 438 to 444 megahertz channel. Also, spacing between the elements is important. For that reason, I will supply both the element lengths and the distance they are apart.

The reflector for our antenna is $13^1/_2$ inches, and the driven element is $6^1/_2$ inches on each side of the boom. Actually, the driven element has to be twice that length as it is folded over. The element is bent at the $6^1/_2$ inch length, then bent back towards the boom. There needs to be a one-half-inch space between the two bent sections, and the second length is also cut at $6^1/_2$ inches.

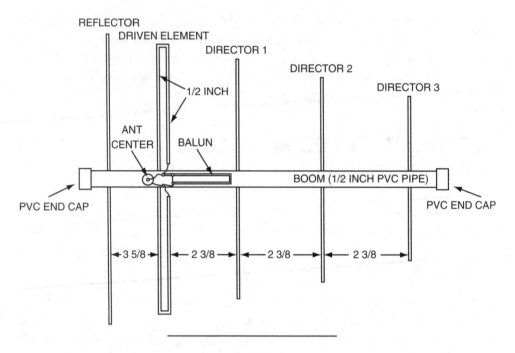

BALUN DETAILS

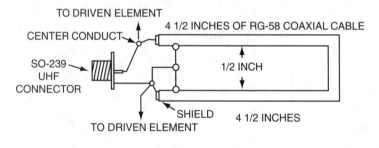

ELEMENT LENGTHS

ALL ELEMENTS ARE MADE FROM 3/16 INCH STEEL ROD

REFLECTOR IS 13 1/2 INCHES END TO END
DRIVEN ELEMENT IS 6 1/2 INCHES FROM BOOM TO END ON EACH SIDE
DIRECTOR 1 IS 12 1/4 INCHES END TO END
DIRECTOR 2 IS 11 3/4 INCHES END TO END
DIRECTOR 3 IS 11 1/4 INCHES END TO END

FIGURE 15.11 Yagi antenna construction diagrams.

The directors for our antenna will be $12^1/_4$ inches, $11^3/_4$ inches, and $11^1/_4$ inches long. Those lengths are in descending order from the driven element. You can see that the antenna gets smaller as it stretches away from the reflector/driven element section.

The boom can be made of just about any nonconducting material. Wooden broom handles are a favorite among some builders, but I prefer to use PVC pipe. In this case it is one-half-inch plastic pipe with corresponding plastic caps on the ends. The caps, in addition to looking tidier, help keep water out of the boom.

The length of the boom is relatively unimportant. It does have to be a little longer than the total length between the reflector and the last director, which in this case is 12 inches. In our design, I made the boom $1^1/_4$ inches longer on each end than the reflector and final director, for a total length of $14^1/_4$ inches. This makes for a very easily managed and unobtrusive antenna.

Drill holes through the boom, making sure to keep them straight, and extend the various elements through the holes. In this design, spacing between the reflector and the driven element is $3^5/_8$ inches. Driven element to first director is $2^3/_8$ inches, and all directors thereafter are $2^3/_8$ inches apart.

One last addition has to be made to our antenna, and that is a *balun*. The balun is present to adjust the antenna impedance to the fifty ohms required by the transmitter. In this design it is a piece of RG-58 coaxial cable. Bend the cable into an elongated loop $4^1/_2$ inches on a side, and solder it to the SO-239 female antenna cable connector.

The outer RG-58 shield, or braid, goes to the outer connection to the SO-239, while the RG-58 center conductor hooks to the SO-239 center connector. With that in place, usually attached along the boom, the antenna now has fifty-ohm impedance and will match the transmitter output.

At this point, you have a highly directional antenna that will vastly improve the carry of any seventy-centimeter transmitter. To adjust driven element size for other frequencies, use the formula 468 divided by the operating frequency in megahertz. For example, if you want to work with a 1240 megahertz frequency, the driven element will be 468 divided by 1240 or 0.37 feet. That works out to a driven element of $4^1/_2$ inches.

As a rule, adding half an inch for the reflector and reducing the directors by a quarter to a half inch as they descend from the driven element is a good starting point. These elements may need some tweaking once the antenna is assembled.

With this antenna, the range of your wireless surveillance system will be greatly enhanced. Also, the chance of someone else picking up the signal is greatly diminished. This is due to the fact that the antenna will direct most of the signal in a rather small pattern. Of course, you have to point the antenna in the right direction—toward the receiver.

VIDEO/AUDIO MODULATOR

This is a simple project (see Figure 15.12) thanks to spare, already assembled and tested, video modulator modules. These modules are plentifully available on the surplus market and can also be found as replacement parts for video games and older computers. You can also often salvage them from damaged or nonfunctional equipment such as VCRs and videodisc players.

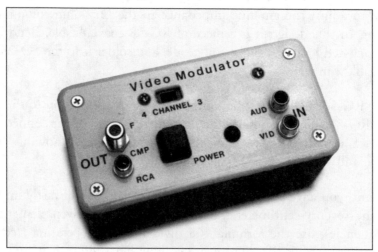

FIGURE 15.12 The completed video/audio modulator project.

I found this one in a surplus catalog, and it was part of an old Commodore computer system. It takes the audio/video input from a camera and turns it into a channel three or four TV signal. When it is used with a standard television, the picture appears on the screen and the sound comes from the speaker. This is a great way to get around the dedicated video/audio output of most video cameras, which is incompatible with standard TVs. Take a look at Chapter 5 for more information about this issue.

As you can see in Figure 15.13, the unit consists of the module, a regulated 12 VDC power supply, and the appropriate input/output jacks. The whole package fits nicely into a 5 x 2$^1/_2$ x 2-inch project case, which makes it very convenient to use. In the version I made for this chapter, I added an On/Off switch with a light emitting diode (LED) ON indicator, but those are optional (see Figure 15.14).

The regulated power supply is of the standard linear design, and it holds input voltages to twelve volts for better stability. I used RCA jacks for the video/audio inputs and an RCA jack and female F connector for the outputs. The F connector for the output is to accommodate the more common F cable arrangement used with television sets. Like all TV signals, this is a composite output.

Labeling of the top cover is very helpful, but not absolutely necessary. I painted the cover a light yellow and marked the various jacks and other controls. With everything in place, I connected the modulator to a TV, introduced a video and audio signal to the inputs, and tested the unit.

With this device at hand, you now have a way to hook up your cameras and other video equipment to any regular television. This can be really helpful in keeping surveillance system costs down or when you want to monitor several different locations.

RADIO FREQUENCY BUG DETECTOR

Here is a device that will allow you to disclose the presence of clandestine equipment monitoring **your** activities in your home and business. Bug detectors do have to use radio transmitters to accomplish their goal, but RF transmission is the most popular method employed by today's bugs.

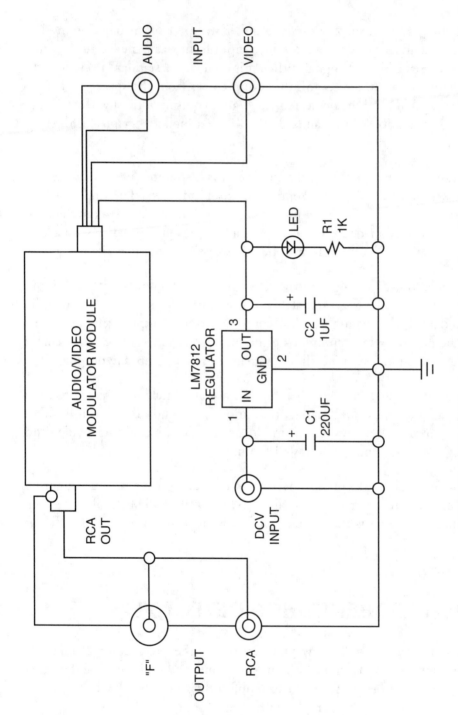

FIGURE 15.13 Video/audio modulator schematic.

FIGURE 15.14   An internal view of the video/audio modulator.

What we have here is a wideband passive detector receiver that will intercept the radio signal that carries the bugging information. It will then display the intensity of that signal on an LED bar graph. This will not only confirm the existence of the bug, but will also lead you to its location. The closer the detector gets to the transmitter, the stronger the bar graph reading.

Looking at the bug detector schematic in Figure 15.15, you can see that the antenna is connected to a coil (L1)/diode (D1) junction. The diode is of the germanium variety, the type that is used for RF detection. Coil L1 provides the required inductance, and the signal is then sent, through gain-control potentiometer R2, to wideband operational amplifier U1.

The op-amp increases the detected signal to a suitable level for the bar graph control chip (U2). In this case the bar graph is managed by an LM3914 integrated circuit (IC). The LM3914 then drives the LEDs in the bar graph display. As the signal becomes stronger, more outputs of the 3914 activate and more

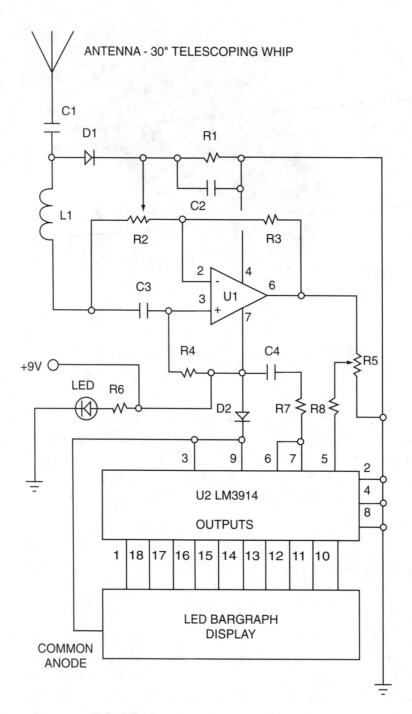

FIGURE 15. Bug detector schematic.

elements of the display light. As you close in on the transmitter, the display keeps you informed.

I am a firm believer in printed circuit boards, but this project can very easily be constructed using perf-board and point-to-point wiring. The type of construction is up to you.

There is nothing particularly critical about the design or layout. Use the normal good habits of electronic construction and you will be fine. I employed IC sockets for U1 and U2; in case one of the chips needs replacing, the sockets make it a whole lot easier.

Any suitable case will work and a telescoping antenna in the twelve to twenty-four inch range, fully extended, is ideal. I suggest telescoping as there will be times, due to signal strength, when you will want to reduce the effectiveness of the detector. An easy way to do that is to shorten the antenna.

Once you have the device installed in its housing and ready to go, it is time to test it. You can use any source of RF radiation to do so, even the wireless TV station we talked about earlier. The FM wireless mics that are so popular are also a great way to put the detector through its paces. Turn the transmitter on and move the bug detector closer to the radiator then farther away. As you move in and out you will see the bar graph change readings.

Next, get someone to hide the transmitter somewhere in the house or building. Using the techniques described in Chapter 8 start hunting the transmitter. It will be difficult at first, but once you pin it down to a specific room, then the process will become more interesting. Hopefully, you will find it entertaining to track down the emitter. Practice a few times with the detector and you will quickly get comfortable with using it.

There is the bug detector. It is a sad commentary on our society that such devices even need to exist. But, be advised they do exist for a very good reason. With this device in hand, you have some protection against the unscrupulous elements of our society.

CONCLUSION

There you have six relatively simple projects to help you with your home or business surveillance system. Each will serve an operational need well, and they really are fairly easy to build. All of this gear can be purchased from video dealers, but that takes some of the fun and pride out of building it yourself. Give at least one of them a try. You might find you like working with electronics. And think of the money you will save.

Have fun!

16

Some Additional Neat Stuff!

Here is a look at some of the neat videotape recorders and liquid crystal displays presently available. I touched on these devices in other areas of the text, but this overview will give you a little more detail concerning two areas that are really movin' and shakin' when it comes to video technology.

VIDEOCASSETTE RECORDING UNITS (VCRS)

Figure 16.1 and Figure 16.2 show how small VCRs have become. Both units are about the size of a cigar box, as can be seen by the presence of the ballpoint pen in the second figure, and weigh in at around twenty-three ounces.

The units employ either digital formatting or "Hi-8" recording to provide excellent horizontal resolution, and their size easily allows them to be carried on a person's body or concealed in a variety of locations.

Portable surveillance systems are ideal for situations where you need a covert, temporary setup. For instance, if you know that someone is stealing from a loading dock, you could set up one of these tiny systems with a motion detector trigger in a parked car outside the dock. Then it's only a matter of time before you capture the thief on tape.

PHOTO COURTESY OF SUPERCIRCUITS, INC. USED BY PERMISSION.

FIGURE 16.1 A tiny VCR.

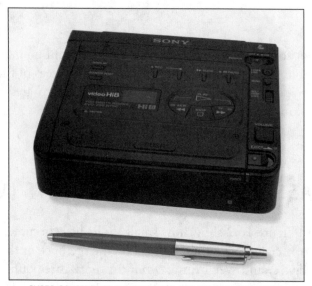

PHOTO COURTESY OF SUPERCIRCUITS, INC. USED BY PERMISSION.

FIGURE 16.2 Really tiny.

If real convenience is required, you will have to go far to beat the micro combination unit shown in Figure 16.3. Here, the small VCR is teamed up with a four inch Thin Film Transistor (TFT) flat-pack LCD for a complete recording/ monitoring system, and it's still the size of a small cigar box.

PHOTO COURTESY OF SUPERCIRCUITS, INC. USED BY PERMISSION.

FIGURE 16.3 A tiny VCR with a tiny monitor.

The price is a bit heavy, almost $1,000 as of this writing, but that will probably drop considerably in coming years. If you really need this combination for a special surveillance job, that need might well justify the cost.

While the monitor/recording assembly shown in Figure 16.4 is offered as part of a portable police in-car system, it is available to the general public. Designed for archiving activity in front of a police vehicle through the windshield, it could be put to a variety of other uses. Just about everything is automatic on this system, but it can also be activated manually.

The system features a high-sensitivity color camera, high-resolution TFT LCD monitor, and either eight-hour or twenty-four-hour recording capability. Price for the eight hour unit is just under $1,500, while the twenty-four hour version rings in at just under $2,500.

PHOTO COURTESY OF SUPERCIRCUITS, INC. USED BY PERMISSION.

FIGURE 16.4   A portable VCR and monitor.

LIQUID CRYSTAL DISPLAY (LCD) MONITORS

As I have remarked before in this book, liquid crystal displays (LCDs) have come a long way in recent years. For starters, the technology used to produce the display matrix has changed considerably. Today, the Thin Film Transistor (TFT) manufacturing approach has become very common and produces excellent picture resolution and quality.

In Figure 16.5, a 5.5-inch TFT LCD unit is shown. The size and versatility of this unit makes it great for a number of applications. These might include space-stingy installations, portable recording arrangements, or even enabling your kids to watch videos in the car.

In Figure 16.6 we have a very portable four-inch TFT LCD unit that, as you can see in Figure 16.7, is often used for checking camera placement. Its size and weight make the device perfect for determining proper camera location and

PHOTO COURTESY OF SUPERCIRCUITS, INC. USED BY PERMISSION.

FIGURE 16.5   A 5.5-inch liquid crystal display monitor..

PHOTO COURTESY OF SUPERCIRCUITS, INC. USED BY PERMISSION.

FIGURE 16.6   A four-inch monitor.

PHOTO COURTESY OF SUPERCIRCUITS, INC. USED BY PERMISSION.

FIGURE 16.7   Very useful and versatile.

camera view adjustment. This can save you hours of running back and forth between your monitor and the camera trying to get the exact angle you need.

That is just one possible use for this gem. Virtually any project that requires a small, self-contained LCD monitor can make good use of this piece of equipment. The unit requires twelve volts direct current (VDC) and is battery operated. It also has RCA jacks for hookup, a tough metal housing for protection, and a socket for attaching the unit to a tripod for extra stability and convenience.

Figure 16.8 shows a really neat combination, although classified as "police restricted" as of this writing. However, I predict similar units will be available to the general public in the not-too-distant future.

This unit combines a four inch TFT LCD and a microwave receiver in either the 1.2 or 2.4 gigahertz range. The result is an extremely compact wireless monitoring station. Thanks to the latest solid-state technology and digital signal control, this unit maintains rock solid performance for both signal reception and picture quality. Watch for something like this real soon in the video surveillance stores and catalogs!

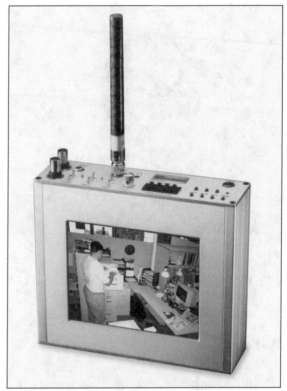

PHOTO COURTESY OF SUPERCIRCUITS, INC. USED BY PERMISSION.

FIGURE 16.8 A wireless LCD monitor.

While this one (Figure 16.9) is a little "Buck Rogers" in nature, it is a version of the video headgear used in virtual reality. Two small (0.7 inch) active matrix LCDs are positioned in this device so the images displayed upon them are seen by the person wearing the gear.

Along with virtual reality applications, this headgear could be used for video surveillance, hazardous waste and bomb disposal purposes, scientific simulation, and more.

No focusing is required, the unit weighs a mere eight ounces, and it operates off six VDC. I am sure you can think of at least one purpose for this gem.

PHOTO COURTESY OF SUPERCIRCUITS, INC. USED BY PERMISSION.

FIGURE 16.9   Wearable LCD monitor.

CONCLUSION

So there you have it. Just some of the "gee-whiz" stuff that can be had today. Just imagine what tomorrow has to offer. As we have seen in this book, technology is red-lining the tachometer these days, and the results will be even more exciting, amazing, and useful video equipment.

PART SIX

Resources

This book doesn't really have enough space to devote much time to explaining many of the acronyms and techno-jargon that litter the user's world. The glossary does its best to condense all the terms you really need to know into one quick stop.

Meanwhile, you may be lucky enough to live somewhere with good shops that sell the kinds of equipment you need. Then again, even if the equipment is available, that doesn't mean that you won't need good advice to go with it. In this section of the book, I provide lists of the places and people to seek out for help, for gadgetry, for information. I hope you'll find it useful.

Glossary

Alternating Current

Non-polarized power that is constantly changing back and forth between positive and negative. Household 120 volt P-P lines are an example.

Amateur Television (ATV)

A section of HAM radio that involves sending and receiving full motion, or fast scan, pictures over amateur radio frequencies.

Amp

Named after the French physicist Andre Marie Ampere, this is the basic unit of current.

Amplifier

A device that increases the power of a signal.

Amplitude Modulation

A form of modulation where the base frequency remains constant but the amplitude, or height, of the individual cycles changes. TV video is AM.

Anode

The positive terminal of a diode, light emitting diode, or battery.

Antenna

Any device that radiates a signal or pulls in a signal.

Audio

The sound portion of a television signal or the output of an amplifier.

Baird, John Logie

A British scientist involved in early 1920s mechanical television.

Balance

This applies to color uniformity or an equilibrium in adjustment.

Beam

In electronics, this is either the stream of electrons generated by a cathode-ray tube, or a directional antenna used to strengthen signals.

Binary

Referring to multiples of 2, and depicted as a string of 0's and 1's. Example: the number 10 would be written as 00001010.

Binary Coded Decimal (BCD)

This is the language of computers based on multiples of 2. Example: 1, 2, 4, 8, 16, etc.

Bit

In this context, this is one piece of the hexadecimal file. A *nibble* (sometimes spelled *nybble*) is made up of four bits, and a *word* is eight bits or more.

Black and White

The original form of television with black, white, and shades of gray.

Breadboard

This term refers to a solderless method of wiring circuits for prototyping. The board has interconnected sockets that the component leads are plugged into.

Capacitor

An electronic component that temporarily stores an electrical charge within a circuit. However, this is not synonymous with battery.

Cathode
The negative terminal of a diode, battery, or other electronic component.

Cathode-Ray Tube (CRT)
This is the electron tube that provides the visual image from a video camera or other television device. A picture tube.

Ceramic Resonator
An electronic component that consists of a quartz crystal and two small capacitors.

Charge-Coupled Device (CCD)
This is a video imaging device used in many modern TV cameras and known for longevity, reliability, and low power consumption.

Chip
This is the slang term for an integrated circuit (IC) and refers to the silicon matrix inside the IC.

Chrominance
The color element of a television signal including red, green, blue, and the color burst information.

Closed-Circuit Television (CCTV)
A secure television system that uses cable to transport the signal as opposed to an RF transmitter.

Coil
Made up of turns of wire, this is an inductive component often used in electronics for filtering and frequency determination.

Color Burst
The NTSC-established frequency of 3.579545 megahertz from which the synchronization signals are derived.

Composite Signal
A television signal that has all the necessary information included on one line. Critical timing phases the information in and out to keep it on track.

Counter

A type of electronic circuit that counts pulses. It can also be used to divide an input signal.

Crooke, Sir William

British scientist who invented the first cathode-ray tube in 1878.

Crystal

Made of quartz, these are used to define a frequency. A thin quartz wafer vibrates when exposed to electricity, and the vibration rate sets the frequency.

Data

This refers to just about any type of information put into, or received from, a circuit using a microcontroller.

Digital

In this context, digital refers to the nature of operation of devices associated with computers or microcontrollers, as opposed to the gray-scale approach used in analog schemes.

Diode

A silicone device that conducts electricity in only one direction, and so changes an alternating current (AC) into a direct current (DC).

DIP Switch

This is a small switch, usually used on a circuit board, that follows the DIP configuration and is compatible with that scheme.

Direct Current (DC)

Voltage that has polarity, that is, a negative and positive side that does not change. Batteries are a good example.

Display

An electronic device, either liquid crystal (LCD) or light emitting diode (LED), that is used to annunciate information.

Dual In-line Pin (DIP)
A standard configuration for one style of IC and for some sockets. It consists of two rows of however many pins or sockets that are used in that particular arrangement.

Electron Gun
The apparatus in a cathode-ray tube that produces the stream of electrons that scans the tube's front screen. Also applies to older image tubes such as orthicons and vidicons.

Farad
Named after Michael Faraday, this is the basic unit of capacitance.

Farnsworth, Philo Taylor
One of the forerunners in the development of modern electronic television. Many consider him the father of television.

Federal Communications Commission (FCC)
The government agency that controls all radio frequency transmissions in the United States.

Field
One of two $262^1/_2$ line scans in the NTSC format. Both scans make up a frame of a television picture. Fields occur every 1/60th of a second.

Frame
Both fields together. Frames occur every 1/30th of a second.

Frequency
The wavelength at which a radio frequency transmitter operates.

Frequency Modulation (FM)
A modulation system in which the amplitude, or height, of the individual cycles stays constant, while the repetition of the cycles changes. The audio signal of television is FM.

Gigahertz

Pronounced *gig-a-hurts* or *jig-a-hurts*, one gigahertz is equal to 1000 mega-hertz. This is within the microwave range.

Henry

Named after American Joseph Henry, this is basic unit of induction.

Hertz

Named in honor of German electrical pioneer Heinrich Hertz, this is the basic unit of frequency. One hertz equals one cycle per second.

Horizontal

In television, having to do with anything that moves back and forth as opposed to up and down.

Horizontal Blanking

Period of time during which the horizontal signal trace is turned off or absent.

Horizontal Pulse

Duration of the horizontal signal, usually about 1.71 microseconds.

Horizontal Sync

The frequency at which horizontal scanning takes place. For NTSC, this is 15,743.26 hertz.

Iconoscope

One of the first image tubes. This one was designed by Vladimir Zworykin in the mid-1920s. A large and relatively light insensitive tube that resembled a glass sauce pan with a bent handle.

Image Dissector

The first successful image tube. Invented by Philo Farnsworth in the early 1920s. It gave birth to electronic television.

Image Orthicon

A vastly improved image tube developed by Vladimir Zworykin in 1945. This tube was, for many years, the television industry standard.

Inductor

Another name for a *coil*, this is a component that consists of turns of wire wrapped around a core that can be metallic, non-conductive, or just plain air.

Infrared (IR) Light

Light that is invisible to the human eye but not to CCD and CMOS cameras. Off the end of the red spectrum, it usually ends at about 750 nanometers.

Input

The signal that is applied to a transmitter, VCR, or amplifier. It can be either video or audio, or both.

Integrated Circuit (IC)

This is a semiconductor device that contains a number of other semiconductors such as diodes and transistors. These are etched into a silicon wafer, and are often a complete circuit within themselves. Access to the circuit is supplied by metal pins that protrude from the case.

Interlaced Scanning

The process of scanning every other line as opposed to each line. Used by the NTSC system to prevent annoying screen flicker and fading.

Internet

This is a network of connected computers that provides great access to information from the private sector, educational institutions, and local, state, and federal government organizations. Very handy!

Jenkins, Charles Francis

An American scientist working on mechanical television in the 1920s.

Kelvin

The kelvin is a temperature measurement that is used to express the color, or temperature, of visible light.

Kinescope

A modification of the Crooke tube, otherwise known as the television picture tube.

Lens

An assembly of optical components, usually made from glass, used to focus light on an imaging device.

Light

The electromagnetic spectrum that starts with infrared and ends with ultra-violet light. The visible portion of this starts at about 710 nanometers (red) and runs to 410 nanometers (violet).

Light Emitting Diode (LED)

These are special-purpose diodes where the gallium/arsenide junction glows when electricity is passed through them. They are also seen as complete seven-segment displays that light up red, green, orange, or yellow.

Liquid Crystal Display (LCD)

A low power form of digital display. These use liquid crystals that twist and turn dark when electricity is applied, thus indicating the information.

Megahertz

One million hertz. This is the unit by which all television frequencies are measured.

Monitor

In this context, this is any device that reproduces the original scene viewed by the imaging device. These can be CRT or LCD based.

Multiplex

In electronics, multiplex refers to connecting elements of a circuit together to conserve input/output port lines. For example, a multiplexed LED display will connect all a, b, and c segments of each individual display unit to a common line, and access each unit with the common anode or common cathode line.

National Television Standards Committee (NTSC)

The group that set the standards for modern electronic television in the United States. It met in the early 1950s and announced its standards in 1953.

Nipkow, Paul Gotlieb

A German scientist who, in 1884, demonstrated the first working mechanical television system.

Ohm

Named after German mathematician George Simon Ohm, this is the basic unit of resistance.

Oscillator

This is the electronics of a device that provides operating frequency. It can be controlled by a quartz crystal, ceramic resonator, resistor/capacitor network, or an external clock source.

Output

In this context, output refers to the video or audio signal from a camera, VCR, or other device.

Perf-Board

This is a type of wiring scheme where the component leads are pushed through evenly spaced predrilled holes and connected with lengths of hookup wire.

Phase Alternating Line (PAL)

The video scanning standard used by Britain and much of Western Europe. Characterized by 625 lines and fifty hertz vertical speed.

Power Supply

The electronic device that supplies the required voltage and amperage to a circuit.

Printed Circuit Board (PCB)

A type of wiring where the connection wires are actually traces of copper metal on a piece of non-conductive base material. Holes are drilled in the base material to run the component leads through, to be soldered on the trace side, or in the case of surface-mount technology (SMT), the traces and components are on the same side of the board, with the components being soldered directly to the traces.

Random Access Memory (RAM)

RAM is a volatile-type memory that is used to store data in a short-term fashion. This memory will lose whatever is in it when power is removed.

Regulator

In this context, a regulator is an electronic device or IC that keeps the operational voltage constant no matter how much the input voltage varies. For microcontroller circuits, this is very important.

Resistor

This is an electronic component that resists electrical flow. It is used to maintain proper bias levels, or as part of an oscillator tank circuit.

Resolution

Applies to the quality of the video display. It can involve the number of scan lines, the number of pixels, or the space between the pixels. Also referred to as dot pitch.

Retrace

Also known as *refresh*, this is the process of the electron beam returning to the top of the screen to begin again.

Sarnoff, David

Chairman of RCA for many years and a strong proponent of electronic television.

Semiconductor

One of a variety of electronic components, such as diodes and transistors. These are at the heart of microcontrollers and get their name from their property of only conducting a certain amount of the total electricity available.

Single In-line Pin (SIP)

A single row of sockets or conductors evenly spaced.

Synchronization

The timing signals and pulses that hold the video signal together. Without sync, the video display would be gibberish.

Systeme Electronique Couleur Avec Memoire (SECAM)
The French version of the NTSC format. This is used in France, French-influenced Africa, the Middle East and much of Eastern Europe. It is characterized by 625 lines and a fifty hertz vertical speed.

Television (TV)
TV is the massive industry of video production and commercial broadcasting. It is also the box that sits in your living room, and the name comes from the Greek *tele* (far) and the Latin *vision* (to see).

Television Camera
Any one of many devices that uses various image sensors to convert an image to an electronic signal.

Transistor
This is a semiconductor device that is used for oscillating, amplifying, and switching. Basically, the potential on the base connection controls the degree of flow within the transistor.

Transmitter
The electronic device that sends video/audio information out over the air via radio frequency energy. Some commercial TV stations use over a million watts of power to accomplish this task.

Ultrahigh Frequency (UHF)
The radio frequency spectrum that runs from about 400 to 900 megahertz.

Ultraviolet (UV) Light
A form of light, with a wavelength just above visible light, that starts at about 410 nanometers and goes down.

Vertical Blanking
The process of shutting off the electron gun during the return trip to the top of the screen.

Vertical Speed
The number of times the screen is scanned in a one-second interval. The NTSC has set vertical speed to sixty fields per second and thirty frames per second.

Very High Frequency (VHF)

The range of frequencies between 30 to 400 megahertz.

Vidicon

The newest and last of the imaging tubes. RCA introduced the device in 1952, and due to its smaller size, higher light sensitivity, and longevity, the tube became an industry standard.

Volt

Named after Italian physicist Count Alessandro Volta, this is the basic unit of electricity.

Westinghouse, George

Founder of his namesake company, George Westinghouse was the first to hire Vladimir Zworykin.

White Balance

The voltage level at which pure white occurs. In the NTSC format, this is 1 volt peak to peak. Most modern color video cameras set this level automatically.

Zworykin, Vladimir Kosma

One of the early researchers in electronic television. He is credited with the invention of the *Iconoscope* and *Image Orthicon* imaging tubes.

Sources

The following is a list of possible sources for materials discussed in this text. I have done business with many of these companies in the past, so I can vouch for them. The ones I have not dealt with were advertised in reliable publications, so I trust them.

I have provided every possible means of contacting these companies, and to the best of my knowledge, all street addresses, phone numbers, and Internet addresses were correct when I was writing this book. If you have Internet access, it is well worth a quick trip to these Web sites to check out the vendors' lines of equipment. You can often request catalogs from Web sites as well as browse and even shop online.

Naturally, this is not every source available, so keep a lookout for new ones. One way to find new sources is to try typing *video*, *CCTV*, *television*, *electronic surveillance*, or any other video-related word into a search engine. You may well be surprised and delighted with what you come up with! Keep in mind that I cannot vouch for the integrity of every vendor you might find through the World Wide Web.

Brick-and-Mortar Sources

All Electronics Corporation
Post Office Box 567
Van Nuys, CA 91408-0567
Phone: (888) 826-5432
FAX: (818) 781-2653
E-Mail: allcorp@allcorp.com
Web: www.allelectronics.com
Good line of surplus parts and equipment, video cameras, and related gear, tools and other electronic stuff.

AllTech Electronics
aka: Computer Circulation Center
2618 Temple Heights
Oceanside, CA 92056
Orders: (888) 404-8848
Phone: (760) 724-2404
FAX: (760) 724-8808
E-Mail: info@allelec.com
Web: www.computerchopper.com
Carries computers, video cameras, digital cameras, and computer components.

Alltronics
PO Box 730
Morgan Hill, CA 95038-0730
Phone: (408) 847-0033
FAX: (408) 847-0133
E-Mail: ejohnson@alltronics.com
Web: www.alltronics.com
Parts, video cameras, and related gear.

American Innovations, Inc.
383 West Route 59
Spring Valley, NY 10977
Phone: (845) 371-0000

FAX: (845) 371-3885
E-Mail: spy@spysite.com
Web: www.spysite.com
Line of wireless surveillance equipment including 2.4 GHz units.

American Science and Surplus

3605 Howard Street
Skokie, IL 60076
Phone: (847) 982-0870
FAX: (800) 934-0722
E-Mail: info@sciplus.com
Web: www.sciplus.com
Tremendous inventory of surplus optics, video gear, electronics, tools, lab ware, scientific equipment, parts, and more.

ATV Research, Inc.

1301 Broadway
Post Office Box 620
Dakota City, NE 68731-0620
Orders: (800) 392-3922
Phone: (402) 987-3771
FAX: (402) 987-3709
E-Mail: atv@pionet.net
Web: www.atvresearch.com
Extensive line of ATV, CCTV, SMATV equipment including cameras, transmitters, receivers, downconverters, antennas, monitors, recorders, and other related equipment.

B.G. Micro

555 N. 5th Street, Suite 125
Garland, TX 75040
Orders: (800) 276-2206
Tech Support: (972) 205-9477
FAX: (972) 205-9417
E-Mail: bgmicro@bgmicro.com
Web: www.bgmicro.com
Parts, video cameras, and peripheral devices.

Black Feather Electronics

4400 S. Robinson Ave
Oklahoma City, OK 73109
Phone: (405) 616-0374
FAX: (405) 616-9603
E-Mail: blkfea@juno.com
Web: www.blkfeather.com
Good line of surplus and new products, electronics, and gadgets.

C & S Sales

150 West Carpenter Avenue
Wheeling, IL 60090
Phone: (800) 292-7711
FAX: (847) 541-9904
E-Mail: info@cs-sales.com
Web: www.cs-sales.com
Good selection of electronic test equipment, kits, tools.

Carl's Electronics

Post Office Box 182
Sterling, MA 01564
Phone: (978) 422-5142
FAX: (978) 422-8574
E-Mail: sales@electronickits.com
Web: www.electronickits.com
Good line of kits, robotics, surveillance equipment, test equipment, and satellite stuff.

CCTV Outlet

1376 N.W. 22nd Ave
Miami, FL 33125
Phone: (800) 323-8746
Fax: (305) 635-3175
E-Mail: sales@cctvco.com
Web: www.cctvco.com
Good line of video cameras and related equipment.

Circuit Specialists, Inc.

220 South Country Club Drive #2
Meza, AZ 85210
Phone: (800) 528-1417
Fax: (480) 464-5824
E-Mail: jr@cir.com
Web: www.web-tronics.com
Parts and video-related equipment.

Contact East, Inc.

335 Willow Street
North Andover, MA 01845-5995
Phone: (800) 225-5370
FAX: (800) 743-8141
E-Mail: sales@contacteast.com
Web: www.contacteast.com
Good inventory of test equipment, tools, lab furniture, and other electronics-related supplies.

DC Electronics

PO Box 3203
Scottsdale, AZ 85271-3203
Order: (800) 467-7736, or (800) 432-0070
Info: (480) 945-7736
FAX: (480) 994-1707
E-Mail: clifton@dckits.com
Web: www.dckits.com
Parts and video transmitters.

Digi-Key Corporation

701 Brooks Avenue South
Thief River Falls, MN 56701-0677
Phone: (800) 344-4539
Fax: (218) 681-3380
E-Mail: webmaster@digikey.com
Web: www.digikey.com
Parts, TV cameras, and video equipment.

Edmund Scientific

60 Pearce Ave
Tonawanda, NY 14150-6711
Phone: (800) 728-6999
FAX: (800) 828-3299
E-Mail: scientifics@edsci.com
Web: www.scientificsonline.com
Good inventory of scientific and electronic parts and equipment, including video cameras and lasers, as well as gear for astronomy, physics, et cetera.

Electro Mavin

2985 East Harcourt Street
Compton, CA 90221
Phone: (800) 421-2442
LA Area: (310) 632-9867
FAX: (310) 632-3557
Web: www.mavin.com
Good line of surplus test equipment, computers, microwave equipment, optics, tools, parts, and gadgets.

Electronic Goldmine

Post Office Box 5408
Scottsdale, AZ 85261
Orders: (800) 445-0697
Phone (Tech): (480) 451-7454
FAX: (480) 661-8259
E-Mail: goldmine-elec@goldmine-elec.com
Web: www.goldmine-elec.com
Surplus part and equipment, video cameras and monitors, tools.

Electronix Express

365 Blair Road
Avenel, NJ 07001
Order: (800) 972-2225
Info: (732) 381-8020
FAX: (732) 381-1572
E-Mail: electron@elexp.com
Web: www.elexp.com
Parts and video equipment.

Home Automation Systems, Inc.

17171 Daimier Street
Irvine, CA 92614-5508
Phone: (800) 762-7846
FAX: (800) 242-7329
Web: www.smarthome.com
Complete line of equipment, devices, and accessories for home security and surveillance. Complete line of X-10 system equipment.

HSC Electronics Supply (Halted Specialty Company)

3500 Ryder Street
Santa Clara, CA 95051
Orders: (800) 442-5833
Phone: (408) 732-1573
FAX: (408) 732-6428
E-Mail: sales@halted.com
Web: www.halted.com
Surplus equipment, video cameras, and related equipment.

Intellicam Systems

8107 Beckett Center Drive
West Chester, OH 45069
Phone: (800) 903-3479
FAX: (513) 942-5521
E-Mail: customerservice@intellicamspy.com
Web: www.intellicamspy.com
Video cameras, transmitters, and related equipment.

Jameco Electronics

1355 Shoreway Road
Belmont, CA 94002-4100
Phone: (800) 831-4242
Cust. Service: (800) 536-4316
Tech Support: (800) 455-6119
Fax: (800) 237-6948
E-Mail: custservice@jameco.com
Web: www.jameco.com
Parts, TV cameras, and video-related equipment.

JDR Microdevices

1850 South 10th Street
San Jose, CA 95112-4108
Order: (800) 538-5000
FAX: (800) 538-5005
E-Mail: sales@jdr.com
Web: www.jdr.com
Parts, TV cameras, and video equipment.

J & R Music World/Computer World

59-50 Queens-Midtown Expressway
Maspeth, NY 11378-9896
Phone: (800) 221-8180
Customer Ser: (800) 426-6027
FAX: (800) 232-4432
E-Mail: custserv@mail.jandr.com
Web: www.jandr.com
Good line of all kinds of electronics, digital cameras, monitors, and computer gear.

Marlin P. Jones and Associates, Inc.

P.O. Box 12685
Lake Park, FL 33403-0685
Orders: (800) 652-6733
FAX: (800) 432-9937
E-Mail: mpja@mpja.com
Web: www.mpja.com
Parts, video cameras, and related equipment.

MATCO, Inc.

2246 North Palmer Drive, Unit 103
Schaumburg, IL 60173
Orders: (800) 719-9605
Tech: (847) 303-9700
FAX: (847) 303-0660
E-Mail: infor@matco.com
Web: www.matco.com
Variety of video cameras and related equipment.

MCM Electronics

650 Congress Park Drive
Centerville, OH 45459-4072
Phone: (800) 543-4330
FAX: (800) 765-6960
Web: www.mcmelectronics.com
Good inventory of general electronic equipment and components with an emphasis on audio, video, and test equipment.

MECI (Mendelson Electronics Company)

340 East First Street
Dayton, OH 45402
Phone: (800) 344-4465
FAX: (800) 344-6324
E-Mail: service@meci.com
Web: www.meci.com
Good line of surplus parts, video gear, and equipment for computers and general electronics.

MEI/Micro Center

411 S. Leap Road
Hilliard, OH 43026
Phone: (800) 382-2390
FAX: (614) 777-2620
E-Mail: csrs@microcenterorder.com
Web: www.microcenterorder.com
Good line of digital cameras, monitors, scanners, and other related equipment.

Micro Warehouse

1720 Oak Street
Post Office Box 3014
Lakewood, NJ 08701-5926
Phone: (800) 367-7080
FAX: (908) 905-5345
Web: www.warehouse.com
Line of digital and analog cameras, scanners, monitors, and more.

Microchip Technology, Inc.

2355 West Chandler Blvd.
Chandler, AZ 85224
Phone: (480) 792-7966
FAX: (480) 792-4338
Web: www.microchip.com
Complete PICmicro Line, PICSTART PLUS Programmer, software, and tech support.

Mouser Electronics

1000 North Main
Manfield, TX 76063-4827
Phone: (800) 346-6873
E-Mail: sales@mouser.com
Web: www.mouser.com
Good source of a wide variety of electronic parts, video equipment, and equipment from ICs to test equipment.

MWK Laser Products

455 West La Cadena #14
Riverside, CA 92501
Phone: (800) 356-7714
FAX: (909) 784-4890
E-Mail: support@mwkindustries.com
Web: mwkindustries.com
Great inventory of lasers, laser-related equipment, and parts.

Parts Express

725 Pleasant Valley Drive
Springboro, OH 45066-1158
Phone: (937) 222-0173, or (800) 338-0531
Fax: (937) 743-1677
E-Mail: sales@parts-express.com
Web: www.parts-express.com
Good inventory of VCR parts, video gear, and maintenance materials.

PC Electronics

2522 Paxson Lane
Arcadia, CA 91007-8537
Phone: (626) 447-4565
FAX: (626) 447-0489
E-Mail: tom@hamtv.com
Web: www.hamtv.com
Extensive line of Amateur Radio Television (ATV) transmitters, exciters, linear amplifiers, receivers, downconverters, antennas, audio boards, and other ATV-related equipment.

PC Mall (Mailing Management Department)

2555 W. 190th Street
Torrance, CA 90504
Phone: (800) 222-2808
FAX: (310) 225-4004
E-Mail: customerservice@pcmall.com
Web: www.pc-mall.com
Good line of digital cameras, monitors, scanners, and related equipment.

Polaris Industries

470 Armour Drive North East
Atlanta, GA 30324-3943
Phone: (800) 752-3571
Fax: (404) 872-1038
E-Mail: sales@polarisusa.com
Web: www.polarisusa.com
Good line of video cameras and related equipment.

Radio Shack and Radio Shack.Com

300 W. Third Street, Suite 1400
Fort Worth, TX 76101
Phone: (800) 813-0087
FAX: (800) 813-0087
Web: www.radioshack.com
Catalog and local store sales of a variety of electronic equipment, parts, video and audio equipment, communications gear, and electronic gadgets and toys.

Ramsey Electronics, Inc.

793 Canning Parkway
Victor, NY 14564
Phone: (800) 446-2295
Fax: (585) 924-4886
E-Mail: help@ramseyelectronics.com
Web: www.ramseyelectronics.com
Video cameras and related equipment, kits, amateur radio equipment, test equipment.

Resources UN-LTD.

300 Bedford Street
Manchester, NH 03101
Phone: (800) 810-4070
FAX: (603) 644-7825
E-Mail: info@RESUNLTD4U.COM
Web: www.resunltd4u.com
Good line of surplus equipment, video cameras, lenses, and related equipment.

RF Parts Company

435 South Pacific Street
San Marcos, CA 92069
Orders: (800) 737-2787
Cust. Ser.: (760) 744-0750
Main Order (Delivery Info): (760) 744-0700
FAX: (760) 744-1943 or (888) 744-1943
E-Mail: rfp@rfparts.com
Web: www.rfparts.com
Complete line of current and obsolete components for all types of radio frequency applications. Tubes, transistors, power modules, rolling inductors, air vane capacitors, and other related parts.

RNJ Electronics, Inc.

202 New Highway
Post Office Box 667
Amityville, NY 11701-0667

Phone: (800) 645-5833 or (631) 226-2700
FAX: (800) 765-3291 or (631) 226-2770
E-Mail: rnjelect@rnjelect.com
Web: www.rnjelect.com
Extensive line of electronic components, video cameras, monitors, related surveillance and security equipment, home/auto and professional audio gear, repair and replacement parts, tools, test equipment, educational materials, kits, communications equipment, microwave video links, and more.

Securetek

7152 South West 47th Street
Miami, FL 33155
Phone: (305) 667-4545
FAX: (305) 667-1744
E-Mail: aseniza@securetek.net
Web: www.securetek.net
Video cameras and related equipment, security equipment.

Spy City (Division of Great Southern Technology)

Post Office Box 1219
Carrolton, GA 30117
Phone: (800) 732-5000
Web: www.spy-city.com
Good line of surveillance and countersurveillance equipment. Catalog also offers some very good information.

Supercircuits

One Supercircuits Plaza
Leander, TX 78641
Phone: (512) 260-0333
Orders: (800) 335-9777
E-Mail: sales@supercircuits.com
Web: www.supercircuits.com
Good line of video cameras, VCRs, transmitters, video links, multicamera systems, and related video surveillance equipment.

The PC Zone

707 South Grady Way
Renton, WA 98055-3233
Phone: (800) 248-9948
Web: www.pczone.com
Good inventory of computer equipment, digital cameras, monitors, scanners, software, and other related items.

Tiger Software

7795 W. Flagler Street Ste 35
Miami, FL 33144
Orders: (800) 888-6111
Phone: (305) 228-5200
FAX: (305) 415-2202
E-Mail: webresponse@tigerdirect.com
Web: www.tigerdirect.com
Full line of computer equipment, digital cameras, monitors, scanners, peripherals, software, and related items.

Timeline, Inc.

2539 West 237th Street, Building F
Torrance, CA 90505
Orders: (800) 872-8878
Tech Info: (310) 784-5488
FAX: (310) 784-7590
E-Mail: mraa@earthlink.net
Web: www.digisys.net/timeline/
Good line of surplus items, including video cameras, CCD imagers, monitors, LCD displays, computer devices and motherboards, and gadgets.

USI Corp

Post Office Box N2052
Melbourne, FL 32902
Phone: (321) 725-1000
Transmitters, telephone recording, surveillance, and countersurveillance.

Index